CRYOGENICS
AND
MEASUREMENT OF PROPERTIES
OF SOLIDS AT LOW TEMPERATURES

CRYOGENICS AND MEASUREMENT OF PROPERTIES OF SOLIDS AT LOW TEMPERATURES

R. SRINIVASAN
Formerly of Raman Research Institute, Bangalore 560080

A.K. RAYCHAUDHURI
S.N. Bose National Centre for Basic Sciences, Kolkata 700098

and

S. KASTHURIRENGAN
Indian Institute of Science, Bangalore 560012

ALLIED PUBLISHERS PRIVATE LIMITED

New Delhi • Mumbai • Kolkata • Lucknow • Chennai
Nagpur • Bangalore • Hyderabad • Ahmedabad

ALLIED PUBLISHERS PRIVATE LIMITED

Regd. Off. : 15 J.N. Heredia Marg, Ballard Estate, Mumbai–400001, Ph.: 022-22626476
E-mail: mumbai.books@alliedpublishers.com
12 Prem Nagar, Ashok Marg, Opp. Indira Bhawan, Lucknow–226001, Ph.: 0522-2614253
E-mail: appltdlko@sify.com
Prarthna Flats (2nd Floor), Navrangpura, Ahmedabad–380009, Ph.: 079-26465916
E-mail: ahmbd.books@alliedpublishers.com
3-2-844/6 & 7 Kachiguda Station Road, Hyderabad–500027, Ph.: 040-24619079
E-mail: hyd.books@alliedpublishers.com
5th Main Road, Gandhinagar, Bangalore–560009, Ph.: 080-22262081
E-mail: bngl.books@alliedpublishers.com
1/13-14 Asaf Ali Road, New Delhi–110002, Ph.: 011-23239001
E-mail: delhi.books@alliedpublishers.com
17 Chittaranjan Avenue, Kolkata–700072, Ph.: 033-22129618
E-mail: cal.books@alliedpublishers.com
81 Hill Road, Ramnagar, Nagpur–440010, Ph.: 0712-2521122
E-mail: ngp.books@alliedpublishers.com
751 Anna Salai, Chennai–600002, Ph.: 044-28523938
E-mail: chennai.books@alliedpublishers.com

Website: www.alliedpublishers.com

ISBN 10 : 81-8424-306-5

ISBN 13 : 978-81-8424-306-2

Published by Sunil Sachdev and printed by Ravi Sachdev at Allied Publishers Pvt. Ltd., Printing Division, A-104 Mayapuri Phase II, New Delhi - 110064

Preface

Though the first helium liquefier was installed in the National Physical Laboratory, New Delhi, as early as the fifties, the growth of low temperature research in India was tardy. Even by the middle of the seventies there were no more than about half a dozen helium liquefiers of low capacity installed in the country. Due to lack of expertise in maintenance the down time of these liquefiers was high.

With the sensational discovery of high temperature superconductivity in 1985 many laboratories in the universities in India wanted to initiate research in this area. The availability of moderately prized closed cycle cryocoolers came as a boon. Today in many research laboratories in the universities in India closed cycle refrigerators are used to carry out simple measurements, mostly of the resistivity, down to 10 K.

Meanwhile research work was initiated down to 0.3 K attainable with a ^{3}He bath cryostat and 30 mK with dilution refrigerator in one or two laboratories in the country. Squid magnetometers were commissioned in about half a dozen laboratories and new larger capacity helium liquefiers were also installed in a few laboratories. But such facilities were mainly created in institutions outside the pale of the university system.

Development work in Cryogenics has been confined to some laboratories in India. Though a few development projects have been successfully completed, commercialization of such developments has not taken place. Among the large scale projects which are currently under way mention may be made of the Superconducting LINAC project at the Inter-University Accelerator Centre in New Delhi, the Superconducting Tokomak project in the Institute for Plasma Research at Gandhinagar, and the superconducting Cyclotron project in the Variable Energy Cyclotron Centre in Kolkata. Low temperature high magnetic field facilities have been set up with financial assistance from the Department of Science and Technology at the Indian Institute of Science in Bangalore and the UGC-DAE Consortium for Scientific Research in Indore.

Measurement techniques have undergone a revolution with the advent of digital signal processing, and automatic data acquisition. However researchers in our institutions of higher education are not completely aware of the capabilities of these techniques. In many universities the only low temperature measurement that is carried out is that of resistivity. There is a need for research scholars to understand how various physical properties of solids can be measured using digital techniques.

This book is aimed at providing this basic knowledge to students and faculty in Indian universities embarking on low temperature research. The authors would like the researcher to build his own cryostat, plan his measurement and obtain results with accuracy achievable with the use of these digital techniques.

For this purpose, we believe that the researcher should have a basic knowledge of the fundamentals of cryogenics, materials properties at low temperature, thermometry and useful principles in the design of cryostats. These are covered in Part I of the book. This part is written by R. Srinivasan and S. Kasthurirengan.

Part II of the book has three chapters. The first two deal with the basic principles of measurement, and digital techniques for the same. Chapter 10 deals with the measurement of various properties of solids, including scanning probe microscopic techniques at low temperatures and noise measurement techniques. One of us (A.K. Raychaudhuri) has written these chapters with insight gained from personal experience.

There is an Appendix dealing with elements of vacuum techniques.

The authors believe that the book can be used in an elective course in low temperature physics at the Master's and pre-Ph.D. levels.

One of the authors (RS) would like to express his indebtedness to Drs. G. and I. Klipping who initiated him in low temperature research and kindled his interest in this area. He is grateful to the Raman Research Institute for providing facilities for preparing this manuscript.

R. Srinivasan
A.K. Raychaudhuri
S. Kasthurirengan

Contents

PART II
TECHNIQUES FOR THE MEASUREMENTS OF PHYSICAL PROPERTIES OF SOLIDS AT LOW TEMPERATURE

PART I FUNDAMENTALS OF CRYOGENICS, MATERIAL PROPERTIES AT LOW TEMPERATURES AND THERMOMETRY

CHAPTER 1

Introduction

The thermal and electronic properties of a material are strongly dependent on temperature. The thermal properties arise due to the vibrational energy levels in a solid. The vibrational energy levels are quantised. The occupational probability of a given vibrational energy level varies with temperature as $\exp(-\varepsilon/k_B T)$. Here ε is the energy of the vibrational level, k_B is the Boltzmann's constant and T is the absolute temperature. Only at a temperature $T < k_B/\varepsilon$ will one see the characteristic behaviour of the thermal properties arising from the quantisation of the energy levels. For example the lattice specific heat of all solids will be proportional to T^3 at very low temperatures as shown by Debye. Such a behaviour will be seen for most of the solids when the temperature is below 4 K.

Electrons in a metal get scattered by lattice vibrations, defects and by other electrons. In a pure metal, lattice vibrations dominate the scattering at high temperature. The mean free path of the electron varies inversely as the temperature. But at very low temperatures residual impurities in a metal dominate the scattering. When we consider multiple scattering at defects, it turns out that there can be constructive and destructive interference occurring between electron waves traveling from one point to another along different paths. Under certain conditions this enhances the probability of the electron to remain at the starting point causing an increase in resistivity of the material as the temperature is reduced. This is called weak localisation and the effect is seen only at low temperatures.

Electrons in a metal can be treated in the first approximation as independent particles moving in an effective potential. However the Coulomb repulsion between the electrons prevents two electrons, whatever their spin orientation, from occupying the same point in space. This correlation in the motion of electrons becomes important in oxide materials and in zero, one and two-dimensional electron gas. The effects of such electron correlations are seen clearly only at low temperatures.

In a magnetic material, the atomic magnetic moments can be lined up in many different fashions at low temperatures. This gives rise to ferromagnetic, antiferromagnetic, ferrimagnetic and spin glass types of behaviour when the temperature of the material is lowered sufficiently.

There are many novel phenomena in condensed matter systems, which are seen only at low temperatures. A large number of Nobel Prizes in Physics have been awarded to significant discoveries at low temperatures.

The above examples provide the justification for studying the behaviour of matter down to very low temperatures.

Low temperature studies started with the liquefaction of air more than a hundred years ago. One by one, the so-called permanent gases were liquefied. Naturally occurring Helium, the gas with the lowest boiling point, was first liquefied by Kammerligh Onnes in 1908. Experiments could now be performed at low temperatures down to 4.2 K, the boiling point of ^{4}He, by immersing the material to be cooled in a bath of liquid helium. Within a few years of the liquefaction of helium, stunning discoveries of superconductivity in mercury and superfluidity in liquid helium were made. This spurred a lot of activity in low temperature physics.

However the activity was confined only to a few laboratories till the invention by Collins of a continuous small-scale helium liquefier, which became commercially available in the nineteen fifties. Any laboratory could install such a liquefier and carry out low temperature measurements relatively easily. Since the invention of the Collins liquefier, further developments in cryogenic engineering took place over a period of fifty years leading to an improvement in heat exchanger technology and the development of reliably operating turbo-expanders. This has helped in increasing the liquefaction capacity, lowering the power consumption and expanding low temperature activity with liquid helium from laboratory scale to an industrial scale. Liquid helium can be bought in India from a bulk supplier.

There was a second development of great significance for physicists involved in low temperature research. This was the development of closed cycle refrigerators to achieve low temperatures. Philips Laboratories at Eindhoven developed the liquid nitrogen machine using a cryogenerator working on a reversed Stirling cycle. This was a machine which could be handled easily in a small-scale laboratory environment and proved to be immensely popular. One of the important applications of this machine in India was to provide liquid nitrogen for the artificial insemination of cattle. The machine has been improved and made modular by incorporating a pressure swing adsorber system to separate nitrogen from the oxygen in air, before the nitrogen gas is liquefied using a Stirling cryogenerator.

Simultaneously there was the development of a closed circuit refrigerator based on the Gifford-McMahon cycle. Using double staging it was possible to produce about 20 W refrigeration at 80 K at the first stage and 1 W refrigeration at 10 K at the second stage. Such refrigerators are manufactured by many companies and are used for cryo-pumping and for laboratory investigations of properties of materials at low temperatures. The need to have a helium liquefier for many low temperature studies was obviated by this development. In India there are many small laboratories with limited budget and the GM refrigerator finds extensive use in such laboratories. A closed circuit refrigerator also finds applications in other areas like cooling IR detector for space applications and for recondensing the evaporating helium from cryostats used in magnetic resonance imaging.

Since the discovery of superconductivity in 1911, physicists strived hard to develop suitable material to wind a superconducting magnet. Such a magnet can produce fields considerably in excess of fields of the order of 1 to 2 Tesla, which can be produced with iron core magnets. They will be light in weight and the power consumption will be negligible even though one passes high currents through the coils. The coils have to be maintained at a temperature much below the critical temperature of the material. The development, after World War II, of Niobium-titanium filamentary wires embedded in high conductivity copper enabled one to reach magnetic fields of 7 to 10 Tesla with the coil cooled to 4.2 K. The development of Nb_3Sn provided a material with a critical temperature of around 18 K and an upper critical magnetic field of around 20 Tesla. One could use a coil of Nb_3Sn within a coil of NbTi to generate magnetic fields up to about 16 Tesla. These superconducting magnets need to be cooled to 4.2 K with a liquid helium bath.

The lowest temperature one could achieve by pumping on a liquid helium bath is 1.2 K. If one uses the very rare, and considerably more expensive, isotope of helium of mass 3, one can reach a temperature down to 0.3 K by pumping on a bath of this isotope. Investigations on a mixture of the two helium isotopes of masses 4 and 3 led to the development of dilution refrigeration. One can achieve continuous refrigeration of a few microwatts at 10 mK using a dilution refrigerator. Dilution refrigerators are commercially available today enabling one to perform experiments down to about 20 milli-Kelvin.

There were further developments, namely Pomeranchuk Cooling and adiabatic nuclear demagnetisation to reach still lower temperatures. But these techniques are used in a few laboratories around the world where expertise in such techniques exist.

The use of high magnetic fields at low temperatures down to a few milli-Kelvin has led to the discovery of fractional quantum Hall effect and the quasi-particle of charge e/3 associated with the fractional quantum Hall state 1/3. The use of Pomeranchuk cooling has led to the discovery of superfluidity in the isotope ^3He.

The discovery of high temperature superconductivity in the cuprate materials in 1985 sparked the hope that one could build superconducting magnets operating at liquid nitrogen temperature and generating high magnetic fields. This hope has not been realised due to the granular nature and high anisotropy in the coherence lengths and conductivity in these materials. However current leads, made of bismuth cuprate, to a conventional super-conducting magnet have been built. If these current leads are maintained at a temperature below 77 K one can pass the high current to the magnet, which is at 4.2 K, through these leads. There will be no power dissipation in the current leads. This has lead to the development of multi-stage GM refrigerators, with the conventional magnet anchored to the last stage at 4.2 K and the warm end of the current lead anchored to the first stage at 80 K. These are cryogen free superconducting magnets and are likely to find more and more use in laboratories involved in the study of materials at high magnetic fields.

To carry out low temperature studies expertise in vacuum technology is essential. All low temperature experiments are carried out in a vacuum of about 10^{-6} millibar. Knowledge of the properties of materials at low temperatures will be of help in the design of cryostats for measuring different properties. One should have appreciation of the techniques of metal joining.

Several different sensors are now available for thermometry. These sensors can be used in different temperature ranges. They have different sensitivities, and times of response in these different temperature ranges. When experiments have to be done under high magnetic fields, one will have to worry about the effect of the magnetic field on the sensitivity of the thermometer.

The use of digital techniques for measurement has improved the measurement process significantly. It is possible to build modular measurement systems using various sources, actuators, and controllers which can be coupled to computers to automate measurements. Significant noise reduction can be achieved by using phase sensitive detection with proper electromagnetic shielding of the set-up. The basic principles of digital measurements, techniques for converting analogue to digital signals, noise reduction, and data acquisition are some of the topics covered in this book. A long chapter on measurement of different physical properties of solids at low temperatures is included.

To a novice low temperature techniques appear difficult. The aim of this book is to provide sufficient basic information, without loading him with too many details, to a research scholar embarking on low temperature studies, and especially working in a small laboratory in India. It is hoped that such information will help him to design his experiment, construct a suitable cryostat for the measurement and to carry out successfully reliable property measurements on solids.

There are several more advanced books dealing with cryogenics. These are listed below for further study.

REFERENCES

1. Randall F. Barron, "Cryogenic Systems", 2nd edition, Oxford University Press, N.Y., 1985.
2. Thomas M. Flynn, "Cryogenic Engineering", Marcel Dekker Inc., N.Y., 1997.
3. Robert C. Richardson and Eric N. Smith, "Experimental techniques in condensed Matter Physics at low temperatures" (Frontiers in physics series, D. Puras ed.), New York, Adison Wesley, 1988.
4. Guy K. White, "Experimental Techniques in low temperature Physics", 3rd edition, Oxford University Press, New York, 1979.
5. K.M. Mendelssohn, "Quest for Absolute Zero", 2nd edition, Taylor and Francis Ltd., London, 1977.
6. Scott, R.B., "Cryogenic Engineering", D. Van Nostrand Co., Inc., Princeton, New Jersey, 1959.
7. A.C. Rose-Innes, "Low Temperature Techniques", The English Universities Press Ltd., London, 1979.
8. G.G. Haselden, "Cryogenic Fundamentals", Academic Press, New York, 1971.

CHAPTER 2

Properties of Cryogenic Gases, Liquefaction and Storage of Cryogenic Fluids

1. INTRODUCTION

In this book we shall confine ourselves to the attainment of a temperature between 120 K and 1.2 K, and the study of properties of solids in this temperature range. Such temperatures are attained either by using liquefied gases or by using closed cycle refrigerators.

At first, a brief description of the properties of the commonly used cryogenic fluids will be given in this chapter. This will be necessary for understanding the liquefaction of gases. This will be followed by a discussion of liquefaction technique and storage of these cryogenic fluids. Closed cycle refrigerators will be discussed in the following chapter.

2. CRITICAL TEMPERATURE, NORMAL BOILING POINT AND LATENT HEAT OF VAPORIZATION

The *equation of state* of a gas is a relation connecting its pressure, volume and temperature. For an ideal gas, in which inter-atomic interactions are absent and the molecules are taken as point objects, the equation of state is given by the *Ideal Gas* law

$$PV = nRT \tag{2.2.1}$$

Here P is the pressure, V is the volume and n is the number of moles of gas. T is the temperature of the gas in the Absolute scale and R is the universal gas constant with a value 8.314 Joule/mole K.

For a real gas, the molecules interact through the van der Waal's attractive force and further the molecules have a finite size. In such a case the equation of state is more complex than the ideal gas law and is described by, say, the *van der Waal's equation* given by,

$$(P + a/V^2)\,(V - b) = nRT \tag{2.2.2}$$

Here the term a/V^2 accounts for the change in pressure due to intermolecular forces and b is the decrease in the volume V, due to the finite size of the gas molecules. It is usual to plot P as a function of V at a constant temperature of the gas. Such a constant temperature plot is known as an *isotherm*.

Figure 2.1 shows a schematic plot of isotherms for a real gas. At high temperatures, since the gas behaves like an ideal gas, the isotherm tends to be similar to the plot PV = constant. However, when the gas temperature is lowered, there exists a given temperature (*Critical temperature* T_c associated with a *Critical pressure* P_c), below which the liquid state of the gas exists. We will discuss about this later. We shall now consider an isotherm below this temperature T_c.

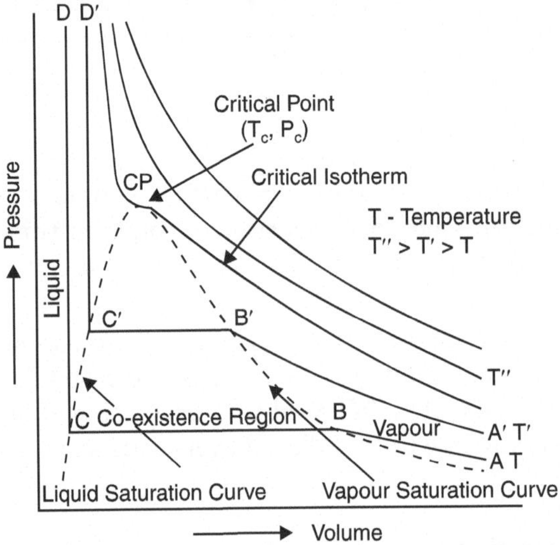

Figure 2.1: Plot of isotherms at different temperatures for a gas. The dotted curve marks the boundary of the two-phase coexistence region. The critical point is indicated in the figure.

When a gas at low pressure (at a point A) is compressed along an isotherm at a low temperature T, the pressure of the gas keeps increasing till the point B is reached. At the pressure corresponding to this point the gas starts to liquefy. If the gas is compressed further, more and more of the gas liquefies at the same pressure. A meniscus is seen separating the gas and liquid phases. At the point C, all the gas has been liquefied. Any further compression results in a sharp increase in pressure and the state of the liquid moves to the point D.

The dotted line CC' up to the *Critical Point* (*CP*) shown in Figure 2.1 is obtained by joining the points at which the gas is fully liquefied on different isotherms. This is called the *liquid saturation curve* (or the *boiling curve*) of the liquid. At a point on this curve, the liquid starts to boil. Similarly if all the points B, B' etc. on different isotherms are joined upto the Critical Point they form the dashed curve *CP-B'-B*. This curve is called the *vapour saturation* (or *condensation*) *curve*. At a point on this curve the gas just starts to condense to a liquid. The portion of the isotherm BC lying within the dotted line *C-C'-CP-B'-B*, is a region in which both the gas and liquid phases co-exist. The temperature of the isotherm is the *boiling point* of the liquid at the pressure P corresponding to the line BC.

The boiling point of a liquid is therefore a function of pressure and increases as the pressure is increased. The boiling and condensation curves are continuous at *CP*. The point *CP* is called the *Critical Point*. At this point the extent of the coexisting region of liquid and vapour phases becomes zero. The liquid and vapour have the same specific volume and the same density. The temperature T_c corresponding to this point is called the Critical temperature, and the pressure at this point is called the Critical pressure P_c.

If we start with the gas at a temperature higher than T_c, the gas cannot be liquefied by isothermally compressing it. The gas can be liquefied by isothermal compression *only* if its temperature is less than T_c. The *normal boiling point*, T_B, of a gas is the boiling point under a pressure of 1 bar. The liquid and vapour phases co-exist along the curve *CB*, as already explained. Heat is absorbed in converting the liquid phase to the vapour phase. This is the *latent heat of vaporization L*. This is an important parameter. Table 2.1 gives the values of T_c, P_c, T_B and *L* for various cryogenic fluids.

Table 2.1: Properties of cryogenic fluids

Fluid	T_{TP} (K)	P_{TP} (mbar)	T_B (K)	T_c (K)	P_c (bar)	L (kJ/litre)
Xe	161.4	820	165.1	289.8	58.9	303
Kr	114.9	710	119.9	209.4	54.9	279
O_2	54. 36	16	90.2	154.3	50.4	245
Ar	83.81	670	87.3	150.9	48.7	224
N_2	63.15	120	77.4	126.0	33.9	160
Ne	24.56	430	27.1	44.5	27.2	110
D_2[(a)]	18.72	170	23.7	38.3	16.6	50
H_2	13.80	70	20.3	33.3	13.0	31.8
^4He	—	—	4.21	5.20	2.28	2.56
^3He	—	—	3.19	3.32	1.16	0.48

[(a)] D_2 is the deuterium molecule formed by atoms of deuterium, which are isotopes of hydrogen and have a mass number 2. The above data are taken from "Matter and Methods at Low Temperatures" by Frank Pobell, Published by Springer Verlag 1992.

3. TRIPLE POINT

One can plot temperature as a function of pressure for the points *B*, *B'* on the condensation curve or the points *C*, *C'* on the boiling curve in Figure 2.1. Since the temperature and pressure are the same at the pair of points *B* and *C*, *B'* and *C'* etc., the plots of *T* versus *P* for both the boiling and condensation curves will fall on a single curve as shown in Fig. 2.2. This is called the boiling curve.

When a liquid is cooled it starts to freeze at a certain temperature and pressure. We have a two-phase region in which the solid and liquid phases coexist. When all the liquid has frozen to form a solid, there is a sharp increase in pressure for further compression. We may plot the temperature and pressure at points on the *freezing curve* (or on the *melting curve* when the solid just begins to melt as the pressure is lowered). The points will again lie on a single curve called the *melting curve*. This is also plotted in Fig. 2.2. The boiling and melting curves of a gas meet at the point *TP* as shown in the figure.

Let us start with a substance in the gas phase at the point A at a high temperature T and lower the temperature at a constant pressure. As long as the pressure is lower than the critical pressure, the gas will condense to form a liquid at the point B.

It remains as a liquid when it is further cooled till it reaches the point C. At this temperature the liquid freezes. Thus the boiling and melting curves divide the p-T space into three regions. To the right of the boiling line is the vapour or gas phase. In the wedge formed between the melting and boiling curves lies the liquid phase. To the left of the melting curve is the solid phase.

Note that the three phases coexist at *TP*, known as the triple point. The temperature at *TP* is called the *triple point temperature* T_{TP} and the corresponding pressure P_{TP} is called the *triple point pressure*.

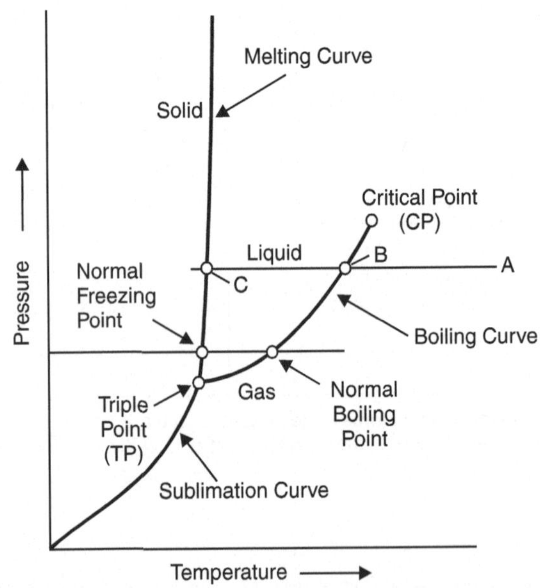

Figure 2.2: Boiling, melting and sublimation curves for a fluid. The critical point is denoted as *CP* and the Triple point as *TP*.

In the P versus T plot shown in Figure 2.2, the boiling curve continues below the triple point as the sublimation curve. This part of the curve separates the solid and vapour phases. On this part of the curve the solid and vapour phases are directly in equilibrium. Table 2.1 gives the values of triple point temperatures and the triple point pressures for different fluids.

A temperature below the normal boiling point T_B can be achieved by pumping on the liquid to lower the pressure on it. With a given cryogen there is a lower limit on the temperature at which it can be used to cool the sample. The cryogenic bath has to be in the liquid phase to facilitate heat transfer by convection. If it goes into the solid phase the heat transfer to the sample will become poor. This requirement sets the triple point temperature as the lower limit at which a cryogenic bath can be used to cool a sample with a cryogen.

Similarly one can attain a temperature above T_B by raising the pressure on the liquid bath. Theoretically, the highest bath temperature one can attain by raising the pressure is the critical temperature T_c. But to attain this temperature one has to apply a pressure of several tens of bars (See Table 2.1 for P_c values).

In practical situations this pressure is limited by the design of the cryostat, especially by its wall thickness. The wall thickness determines the heat leak into the cryogenic bath from the top of the cryostat. To keep this low, one uses low thermal conductivity material such as stainless steel for construction for the cryostat. The wall thickness is chosen such that the upper limit to the pressure that can be safely withstood by the wall is around a few pounds per square inch above atmospheric pressure. The highest temperature to which the bath can be used is therefore the boiling point of the liquid at this safe pressure.

Except helium all gases have a triple point at which the solid, liquid and vapour phases coexist. Helium will not become a solid at a pressure lower than 25 bars. *The melting and boiling lines do not intersect and so helium does not have a triple point in the above sense.*

The phase diagram for liquid ^4He is shown in Figure 2.3. Helium has two isotopes. ^4He has a mass number 4 and an isotopic abundance of 99.7%. The rare isotope ^3He has a mass number 3. It has to be produced in a nuclear reactor and is very expensive. When the word "helium" is used, it normally refers to the isotope ^4He. Helium undergoes a different type of phase transition in the liquid phase. When the boiling point of liquid helium reaches a temperature of ~2.17 K at a pressure of ~50 mbar, the properties of liquid helium undergo a dramatic change. Liquid helium changes from phase I (its normal phase) to phase II (Superfluid phase) at this temperature. Liquid He II has a very high thermal conductivity compared to liquid helium I and it can also flow through narrow channels without any viscosity as long as the velocity of flow is below a critical value.

The transition from phase I to phase II is called the *Lambda transition*. The lambda transition temperature, T_λ, is a function of pressure and this transition line meets the boiling line of liquid helium at a point λP. This is a triple point at which liquid helium I, liquid helium II and vapour helium are in equilibrium.

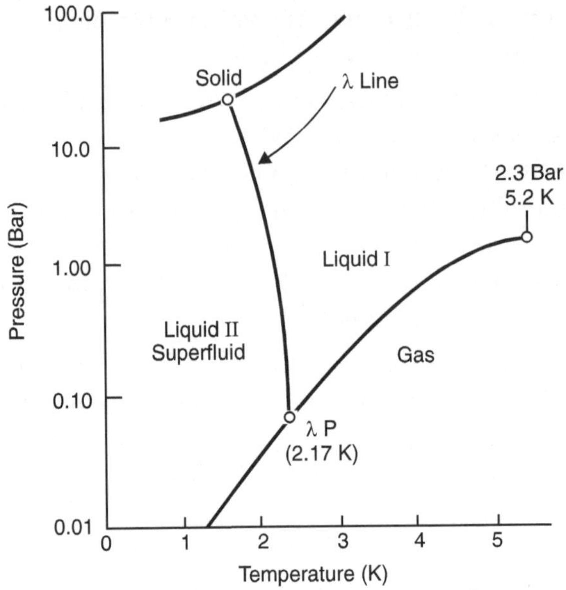

Figure 2.3: Phase diagram for ^{4}He. The solid, liquid and vapour phases in ^{4}He do not coexist. However Liquid I (normal), Liquid II (superfluid) and solid phases coexist.

4. THERMODYNAMIC PROPERTIES OF CRYOGENIC FLUIDS

The important thermodynamic properties for any fluid are *Internal energy U*, *Enthalpy H* and *Entropy S*. The internal energy is made up of the kinetic energy of molecular motion in the fluid and the potential energy due to inter-molecular interaction. The internal energy of a material increases as the temperature increases. When the temperature is raised at a constant volume of the material, the material does no external work. The rate of change of internal energy with temperature of a gram mole of the material at constant volume is called the molar specific heat, C_V, of the material.

$$C_V = (\partial U/\partial T)_V \tag{2.4.1}$$

where U is the internal energy per mole[1].

The next important thermodynamic quantity is the enthalpy defined by

$$H = U + PV \tag{2.4.2}$$

The specific heat at constant pressure is defined by

$$C_p = (\partial H/\partial T)_p \tag{2.4.3}$$

[1] Henceforth, whenever we discuss about the thermodynamic properties it will be understood that we refer to the value of the property for a gram mole of the material if capital letter is used for the symbol. If a small letter is used the value usually refers to 1 g of the material.

In an actual measurement of the specific heat it is more convenient to keep the pressure constant rather than the volume. There are two contributions to C_p, the first term arising from the change in the internal energy of the fluid, and the second term from the external work performed by the fluid in changing its volume at constant pressure. So C_p is larger than C_V. For a perfect gas

$$C_p - C_V = R \tag{2.4.4}$$

For any material

$$C_p - C_V = TV\beta^2/\chi \tag{2.4.5}$$

where T is the temperature, V is the molar volume, β is the coefficient of volume expansion and χ is the isothermal compressibility.

The volume expansion coefficient is defined by

$$\beta = (1/V)\,(\partial V/\partial T)_p \tag{2.4.6}$$

and the isothermal compressibility χ of the material is defined by,

$$\chi = -(1/V)\,(\partial V/\partial P)_T \tag{2.4.7}$$

Using this relation and the measured values of C_p, β and χ one can deduce the value of C_V. Figure 2.4 shows the variation of C_p with temperature for some cryogenic fluids.

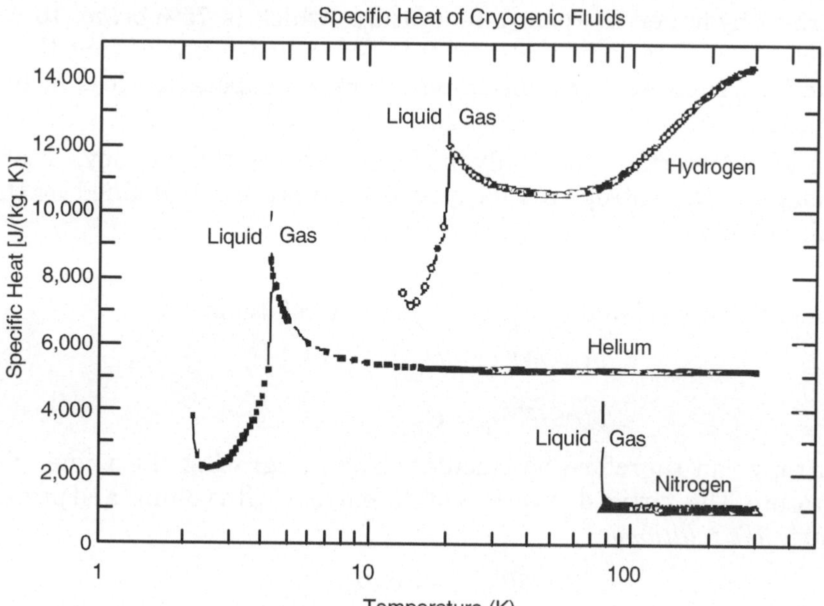

Figure 2.4: Specific heat as a function of temperature of some cryogenic liquids (calculated using a program developed by NIST, Boulder, Colorado)

The data were calculated using a program developed by NIST, Boulder. For liquid helium the specific heat shows an anomalous behaviour at the Lambda transition temperature. The variation of C_p near $T_{\lambda P}$ resembles the inverted Greek letter *Lambda*. That is why the transition is called the Lambda transition.

In the case of the hydrogen molecule we have the following situation. The two electrons responsible for the binding of the hydrogen molecule have opposite spins. But the spins of the two protons can be either parallel or anti-parallel. In the former case, the total nuclear spin I of the molecule is 1 and in the latter case the total nuclear spin I is zero. The energy states in the former case are triplets $(2I + 1 = 3)$. In the latter case, the states are singlets. A transition between a triplet and a singlet state is forbidden in the first order.

Molecules in the triplet state are called *ortho* hydrogen and molecules in the singlet state are called *para* hydrogen. At temperatures at which thermal energy $k_B T$ is high compared to the energy level difference between *ortho* and *para* states, (for example 300 K), equilibrium hydrogen has thrice as many molecules in the ortho state as in the para state. At the temperature of liquid hydrogen, in thermal equilibrium, almost all the molecules are in the *para* state.

When hydrogen is liquefied, the molecules in the *ortho* state at room temperature take a long time to come to the *para* state because the transition is forbidden in the first order. Such a transition also results in a release of heat, which causes an excessive evaporation of the liquid hydrogen. However a catalyst such as nickel speeds up the process of conversion. Such a catalyst is used to convert the equilibrium hydrogen at room temperature (which is 75% *ortho*) to equilibrium hydrogen at 20 K (which is almost fully *para* hydrogen) during the process of liquefaction. Then during subsequent storage evaporative loss of hydrogen is minimized.

The next important thermodynamic property is the entropy, S, of the fluid. The change in the entropy dS for a reversible process is defined by the thermodynamic relation

$$TdS = dU + PdV \tag{2.4.8}$$

The entropy is related to C_V and C_P by the relations

$$T(\partial S/\partial T)_V = C_V \tag{2.4.9}$$

and

$$T(\partial S/\partial T)_P = C_P \tag{2.4.10}$$

Entropy can therefore be calculated by integrating the value of C_P/T over temperature. The partial derivatives of the entropy with volume and pressure satisfy the *Maxwell's relations*;

$$(\partial S/\partial P)_T = -(\partial V/\partial T)_P \tag{2.4.11}$$

$$(\partial S/\partial V)_T = (\partial P/\partial T)_V \tag{2.4.12}$$

It is convenient to plot *isobars* (curves at constant pressure) in a *T-S diagram*. Such a diagram is useful for the study of liquefaction cycles. A process in which entropy does not change is called an *isentropic* process or *reversible adiabatic* process. On the *T-S* diagram, this process is represented by a line parallel to the temperature axis. An isothermal process in which the temperature remains constant is represented by a line parallel to the entropy axis.

The physical interpretation of entropy is that it is a measure of the *disorder* present in the system. The entropy is higher in the gas phase than in the liquid phase. Similarly, it is higher in the liquid phase than in the solid phase. The compression of a gas reduces the phase space available for the gas molecules and so results in a decrease in entropy. Another example is a paramagnetic system. In the paramagnetic phase individual atomic magnetic moments are aligned at random and the phase has a higher entropy than the ordered phase in which the magnetic moments are forced to align in the direction of an applied magnetic field.

When a transition takes place from a liquid to vapour phase, the temperature remains constant, but the entropy increases. The *latent heat of vaporization* is therefore

$$L = T_B(S_{Vap} - S_{Liq}) \tag{2.4.13}$$

A similar relation exists for the latent heat of fusion when a solid melts to form a liquid.

5. LIQUEFACTION OF GASES

Since this is not a book on cryogenic engineering we will not deal in any great detail with the topic of liquefaction of gases. However the principles used in liquefaction will be briefly summarized. The critical temperatures of most of the gases listed in Table 2.1 are below room temperature. In order to liquefy a gas it has to be cooled below the critical temperature.

A gas can be cooled by using one of two following processes (*a*) *adiabatic expansion* and (*b*) *isenthalpic expansion*.

5(a) Adiabatic Expansion

In adiabatic expansion the gas is expanded against an external force so that it does work on the external force. At the same time no heat is allowed to enter or flow out of the gas. If such an expansion is carried out infinitesimally slowly[2] then the process will be reversible and there will be no entropy change in this process. That is why this process is called isentropic or adiabatic expansion.

According to the first law of thermodynamics the internal energy of the gas must decrease by an amount equal to the external work done. Since the internal

---

[2]When the gas is neither expanding nor contracting, the force exerted by the gas on the surroundings is balanced by an external force exerted by the surroundings on the gas. If the force exerted by the gas is infinitesimally larger than the external force then the gas expands slowly pushing the piston out. There is no accelerated motion of the piston.

energy of the gas is a function of temperature and is higher the higher the temperature, adiabatic expansion will cool the gas.

Adiabatic expansion can be realised by enclosing the gas in a cylinder provided with a snugly fitting piston. This is schematically shown in Fig. 2.5. During the upward stroke of the piston, the inlet valve is open and gas at high pressure is admitted into the cylinder (Fig. 2.5(*a*)). After the piston has moved a part of its way, the inlet valve is closed (Fig. 2.5(*b*)). The continued upward motion of the piston, due to rotation of the flywheel to which it is connected, causes the gas to expand and cool (Fig. 2.5(*c*)). When the piston has reached the top of its stroke, the outlet valve opens. The piston in its downward motion pushes the cold gas out at low pressure (Fig. 2.5(*d*)). The outlet valve closes after all the gas has been pushed out and the piston is at the end of its downward stroke. The cycle of operations is repeated in each period of motion of the piston.

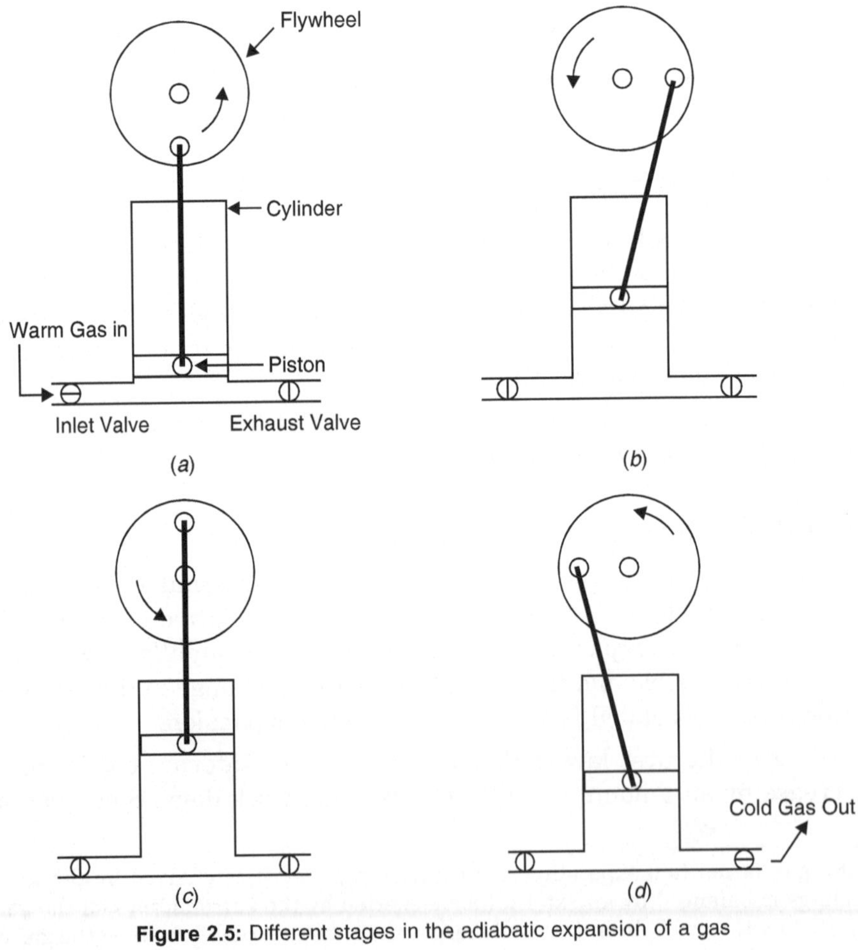

Figure 2.5: Different stages in the adiabatic expansion of a gas

One can also use a turbine expander. The expanding gas sets the turbine in rotation at high speed. The turbine is connected to a brake motor and the external work is dissipated in the brake motor. A turbine expander is always used in high capacity liquefiers. Modern moderate capacity liquefiers also use turbines because they are more efficient than piston cylinder type expansion engines. *Adiabatic expansion always produces cooling in any gas.*

5(b) Joule-Thomson Expansion

The second method of cooling is by *Isenthalpic expansion*. In this method, a gas at high pressure is expanded through an orifice or a porous plug. This is called *Joule-Thomson (JT) expansion*. No heat enters or leaves the system. The enthalpy H of a gas defined in equation (2.4.2) is held constant in Joule Thomson expansion, but the contributory factors *U* and *PV* change. If the product *PV* decreases on going from the high-pressure side of the orifice to the low-pressure side, then the internal energy *U* of the gas will increase after expansion. This implies that the gas will get hot. On the other hand, if the product *PV* increases on going from the high pressure to the low-pressure side of the orifice, the internal energy will decrease on expansion and the gas will be cooled. How the product *PV* will behave on changing the pressure will depend on the temperature of the inlet gas.

Thus *JT* expansion may produce either *heating* or *cooling* of the gas depending on the initial temperature of the gas for a given operating condition. The temperature at which there is no change in temperature of the gas on isenthalpic expansion is called the *inversion temperature*. This temperature is a function of inlet pressure. There is a *maximum inversion temperature* for any gas. *JT* expansion will produce heating if the temperature of the gas is more than the inversion temperature. On the contrary, if the temperature of the gas is lower than the inversion temperature at the inlet pressure, *JT* expansion will produce cooling. For Neon, Hydrogen and Helium the maximum inversion temperature is below room temperature. **Joule-Thomson expansion at room temperature will produce heating in these gases, whatever be the inlet pressure.** These gases will have to be cooled below their maximum inversion temperature for JT expansion to produce cooling.

6. HEAT EXCHANGERS

Usually in a liquefier the pressure on the liquefied gas will be close to the atmospheric pressure. Since the normal boiling point of all the gases is considerably below room temperature, the gas must undergo progressive cooling over a period of time for it to be liquefied. This is done by using a heat exchanger. The principle of *iterative cooling* is illustrated in figure 2.6. In this recuperative heat exchanger, there are two paths for the flow of the gas. The high-pressure stream flows through one path and the low-pressure stream flows in the opposite direction in the other path.

To obtain iterative cooling compressed gas at room temperature flows through the high-pressure arm of the heat exchanger. A part or whole of the gas undergoes either adiabatic or *JT* expansion which produces cold gas at low pressure. This cold gas then passes through the low-pressure path in the heat exchanger in counter-flow. It cools the incoming high-pressure stream and enters as room temperature gas at the inlet of the compressor. It is compressed again and undergoes this cyclic operation. The gas coming out of the high-pressure path of the heat exchanger gets cooled progressively and reaches a temperature much below the inversion temperature.

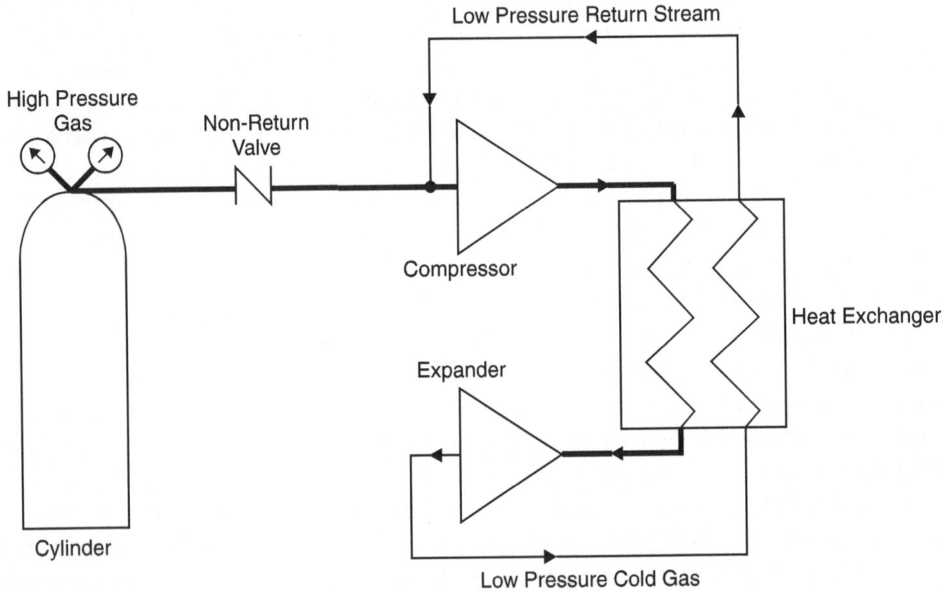

Figure 2.6: Iterative Cooling using a recuperative heat exchanger

Important considerations in the design of a heat exchanger are:

(a) *Heat transfer efficiency:* This depends on the area over which heat is transferred from the warm to the cold streams, the velocity and nature of flow (turbulent or laminar) and the properties of the gas.

(b) *Pressure drop:* A pressure gradient is necessary to drive the flow of gas through the heat exchanger. The pressure on the upstream side must be more than the pressure on the down-stream side. One would like to keep this pressure drop as a small fraction of the average pressure of the flowing gas. This imposes a constraint, which is less serious in the high-pressure stream and more serious in the low-pressure stream. The pressure drop depends on the area of cross section of the tube, the length of the flow path, the nature of the flow and the properties of the gas.

(c) *Thermal capacity:* The thermal capacity of the heat exchanger is a product of its mass and specific heat. In the cool down process the heat exchanger is cooled by several tens to a few hundreds of degrees K. The cool down time will be longer the higher the mass of the heat exchanger. The mass will depend on the density of the material used, wall thickness of the up and down stream channels, and length of path. The design and choice of material of construction will be such as to minimise the mass of the heat exchanger consistent with the other two conditions mentioned above.

Any liquefier uses a train of heat exchangers in series. There are many designs for a heat exchanger to improve heat transfer efficiency. The design to be adopted in a liquefier depends on the mass flow of the gas and convenience of fabrication. In small liquefiers the *tube-in-shell* type of heat exchangers may be preferred. But in large capacity liquefiers, brazed aluminium heat exchangers are used. Further information on heat exchangers will be found in the appropriate references given at the end of this chapter.

7. SCHEMATIC DIAGRAM OF A HELIUM LIQUEFIER

Figure 2.7 shows a schematic diagram of a helium liquefier. Various points are indicated by numbers on this diagram. Figure 2.8 shows the different states in the thermodynamic cycle on a T-S diagram. The numbers in the T-S diagram correspond to the points carrying the same numbers in Fig. 2.7.

Pure helium gas of 99.995% purity is compressed in a water-cooled compressor *C* to a pressure of 16 bars (Points marked 0 to 1 in Fig. 2.8). The heat of compression is removed by cooling water. Special attention must be paid to the removal of oil vapour from the compressed gas stream as this will clog the flow paths in the colder regions of the liquefier. The high-pressure stream (indicated by thick lines in Fig. 2.7) passes through a heat exchanger H_0 and is cooled by flowing liquid nitrogen. This process is shown as the segment 1 to 2 on the high-pressure isobar in Figure 2.8. The gas then passes through the first (H_1) of a train of five heat exchangers. At the exit of H_1 (point 3 in Figure 2.8) nearly 50% of the gas is taken to an expansion engine E_1 (a turbo-expander in Fig. 2.7).

The rest of the high-pressure gas goes through a second heat exchanger H_2. This is shown by the segment 3-4-5 in Fig. 2.8. A tube *AD*1 containing adsorbent charcoal is placed before the inlet of the expander E_1 so that any residual water or oil vapour will be adsorbed by the charcoal. The high-pressure gas expands through E_1 to a pressure slightly above 1 bar. This expansion, which in practice is not exactly isentropic, is shown by the segment 3-13 in Figure 2.8. The gas after expansion enters the low-pressure side of H_2 and returns to the compressor. The high-pressure gas coming out of H_2 passes through a third heat exchanger H_3. About 30% of the high pressure stream coming out of H_3 enters a second expander E_2 at point 5. This expander is also provided with a second tube *AD*2 containing adsorbent charcoal. The rest of the high-pressure stream flows through the heat exchanger H_4 (segment 5-6 in Fig. 2.8).

The gas passing through E_2 expands to a pressure slightly above 1 bar and returns to the low-pressure side of heat exchanger H_4 (segment 5-11 in Fig. 2.8). The high-pressure stream from H_4 enters a final heat exchanger H_5 (segment 6-7) and then expands through a needle valve *JT* to a pressure slightly above 1 bar.

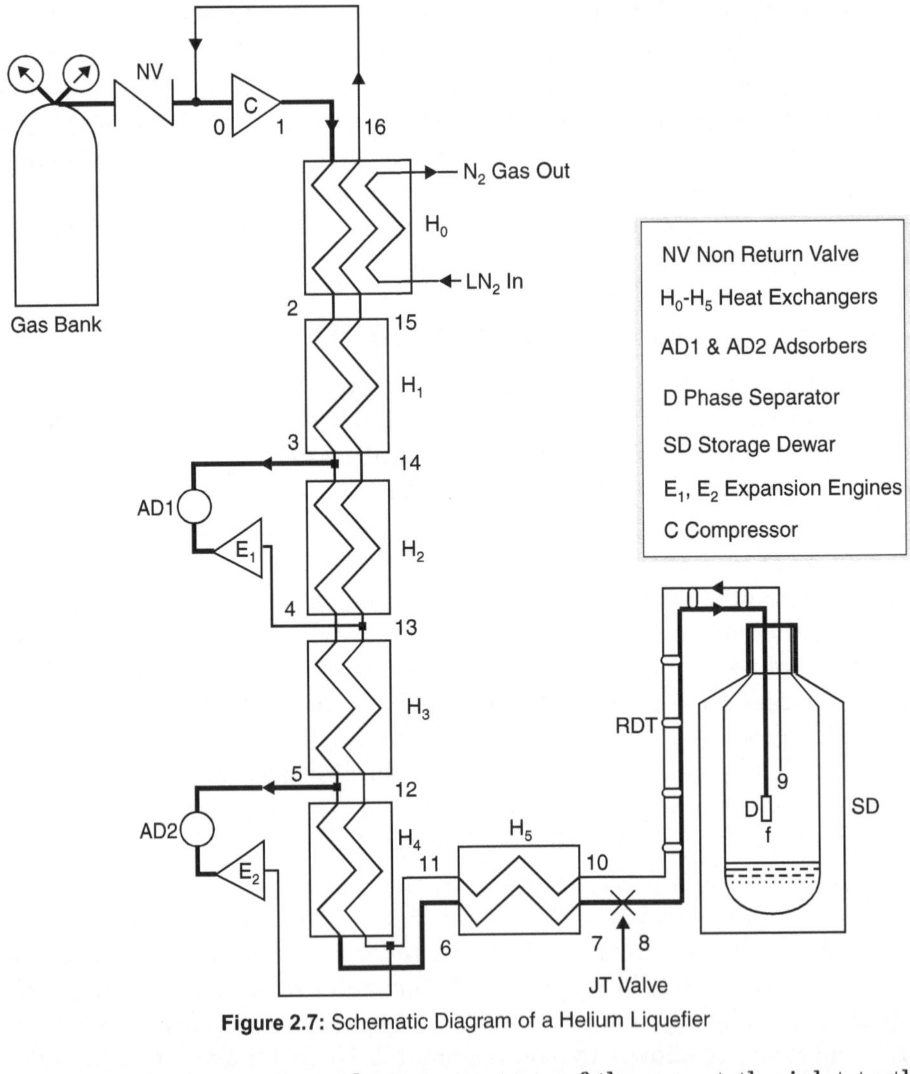

Figure 2.7: Schematic Diagram of a Helium Liquefier

When the dewar is warm, the temperature of the gas at the inlet to the *JT* valve falls below the inversion temperature several hours after the start and the *JT* expansion cools the gas. When the temperature drops to the boiling point of the liquid, part of the gas is liquefied. This gas liquid mixture then passes through the remote delivery tube *RDT* connecting the liquefier to the storage dewar *SD*. The diffuser *D* at the end of the *RDT* ensures the phase separation of the liquid and the gas. Liquid helium is collected in the storage dewar, while the cold gas is returned through the low-pressure side of the remote delivery tube to point 10 of heat exchanger H_5.

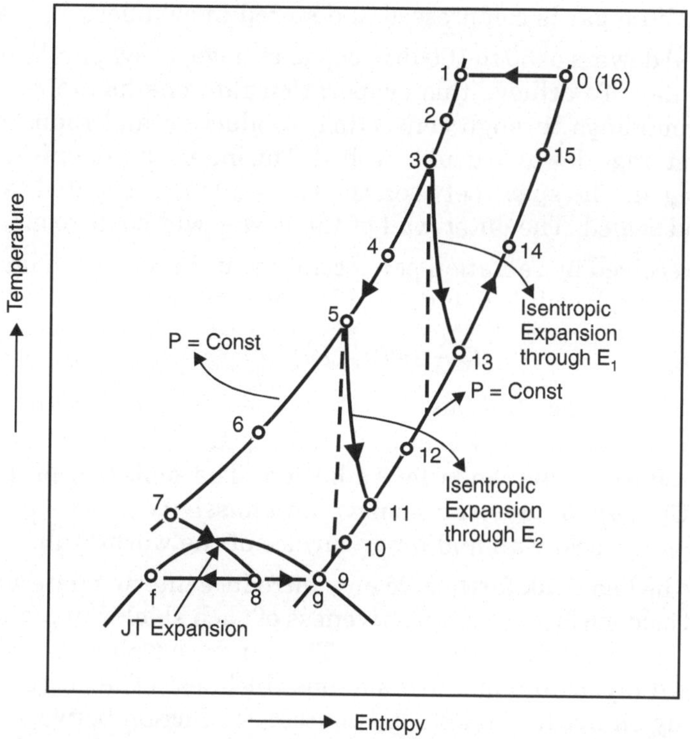

Figure 2.8: T-S diagram for helium showing the thermodynamic cycle for liquefaction

The first time the liquefier is commissioned the storage dewar is warm. It takes several hours for liquid helium to collect in the dewar. Even after collection it takes a day or two before the temperature distribution inside the dewar reaches a steady value and the evaporation of liquid helium reaches a steady state. If the storage dewar is not allowed to warm up, subsequent runs of the liquifier take only a couple of hours before liquid helium starts collecting in the dewar.

Nearly 10 to 15% of the total mass flow of helium from the compressor will be converted to liquid. To make up for this mass that is liquefied the compressor will draw additional gas from the gas bank. Of course, the liquefier will also contain many valves, regulators, temperature and pressure indicators etc which are not shown in the figure. Many safety interlocks are also provided to protect the liquefier from faulty operation.

8. STORAGE OF LIQUEFIED GASES

The liquefied gas is at a temperature much below the ambient temperature and at a pressure slightly more than the ambient pressure. Heat will leak from the ambient temperature of the surroundings to the cryogen. This heat will make the liquid cryogen to boil and evaporate. The evaporating gas is usually let out to the atmosphere. But if it is helium, which is expensive in a country like India, the evaporating helium gas is compressed and stored in cylinders for reuse.

Commercial dewars of 30 to 100 litre capacity have an evaporation rate between 1% to 3% per day. To achieve this evaporation rate one has to reduce heat leak from the surroundings through convection, conduction and radiation. To reduce heat leak, the storage dewar is double walled. The inner container is used for storing the liquid cryogen. The space between the two walls is evacuated to a pressure of 10^{-6} mbars and sealed. The outer wall of the dewar will be at room temperature.

The heat received by radiation per second by unit area of the inner wall of the dewar from the outer wall is determined by the equation

$$Q = \varepsilon\sigma(T_h^4 - T_c^4) \tag{2.8.1}$$

Here σ is the Stefan's constant and has the value 5.67×10^{-8} W/(m^2 K^4). T_h and T_c are the temperatures of the warm and cold surfaces respectively and 'ε' is the effective emissivity of the walls for radiation. The emissive properties will be discussed in Chapter 5. One can achieve an emissivity of 0.01 by polishing the outer surface of the cold wall and inner surface of the warm wall.

To reduce the heat leak further we may introduce intermediate shields between the warm and cold surfaces. The effectiveness of such shields in reducing radiation heat transfer is discussed in Chapter 5. The inner container is wrapped around with aluminised mylar film of a few microns thickness. The successive layers are separated by nylon net to reduce solid contact conduction between one layer and the next. Holes are punched in the film to provide for a path for the gas between the layers to be removed by pumping. The number of layers to be wound and the layer density are determined by the competition between radiation heat leak and heat leak through solid contact conduction. As the number of layers increases the radiation heat leak decreases. To make the winding compact one would like to have as many layers per centimetre thickness as possible. As the layer density increases, heat transfer through solid contact conduction increases. This ultimately limits the layer density. At the top the layers are connected to rings brazed on to the long neck of the dewar at different points. This is shown in Figure 2.9.

During the initial cool down of the dewar, the evaporating vapour cools the layers to different temperatures in a few hours through these thermal contacts. It takes 24 to 48 hours, depending on the size of the dewar, to obtain an equilibrium temperature distribution in the insulating layers when the dewar is filled for the first time. As long as the dewar is kept cooled and is not allowed to warm up the equilibrium temperature distribution between the different layers remains little disturbed.

Another contribution to the heat leaking into the cryogen comes through heat conduction from the top of the dewar through its neck. To reduce this heat conduction, one chooses a material that has a low thermal conductivity and high mechanical strength. The material commonly used is stainless steel 304. The high mechanical strength of this material allows us to choose a small wall thickness (about 0.2 to 0.5 mm) for the inner container. One should remember that the temperature at one end of the container is nearly 300 K while at the other end it is 4.2 K if it contains liquid helium. The thermal conductivity of any solid material is a function of temperature. The variation of the thermal conductivity of different materials as a function of temperature will be discussed in Chapter 5. The calculation of heat leak in such a case will be illustrated there.

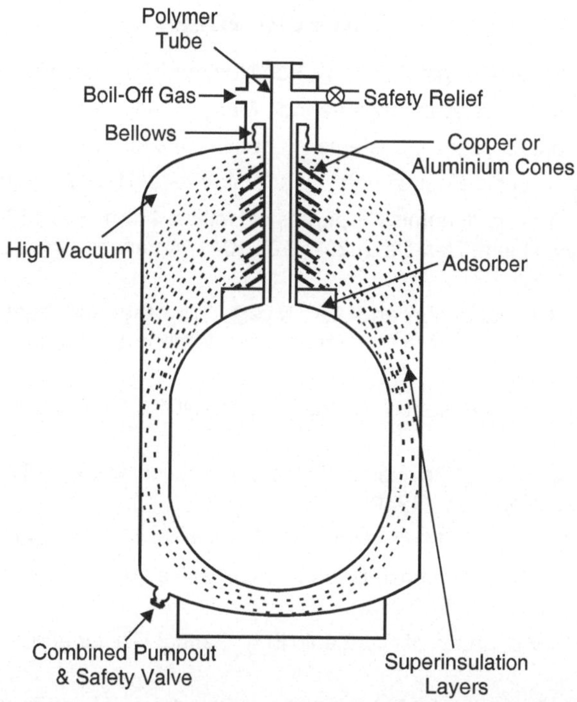

Figure 2.9: A schematic diagram of a storage dewar for liquid helium

A large capacity storage dewar for liquid helium will have a port for the inlet of the remote delivery tube from the helium liquefier, another port for a transfer tube to transfer the stored liquid to a transportable container, a pressure gauge to read the pressure, and a liquid level gauge. There will be a pressure relief valve to vent the vapour to the atmosphere if the pressure in the dewar exceeds a set value. There will be a side port for re-evacuation and sealing of the space between the two containers. In the unlikely event of the failure of the vacuum when the dewar contains liquid cryogen, the pressure in the dewar will increase rapidly due to the quick boil-off of the liquid. To prevent damage to the dewar due to such a rapid pressure rise, there will be a bursting disk fitted to the top of the dewar. This disk will burst and release the cold gas to the atmosphere. Figure 2.9 shows a schematic diagram of a liquid helium storage dewar. It can be seen that the evaporating vapour cools the radiation hats and they in turn cool the multilayer shields.

Large laboratory storage dewars of 500 litre to 1000 litre capacity for liquid nitrogen usually come in a horizontal version. To transfer liquid nitrogen from these dewars to transportable containers a self-pressurising arrangement is used. Nowadays in large cities in India liquid nitrogen is transported in tankers and filled into huge vertical storage tanks of a few cubic metres capacity at the place of use. These tanks are insulated using powder insulation between the inner and outer walls.

REFERENCES

1. Scott, R.B., "Cryogenic Engineering", D. Van Nostrand Co., Inc., Princeton, New Jersey, 1959.
2. Thomas M. Flynn, "Cryogenic Engineering", Marcel Dekker Inc., N.Y., 1997.
3. Randall F. Barron, "Cryogenic Systems", 2nd edition, Oxford University Press, New York, 1985.
4. B.A. Hands, "Cryogenic Engineering", Academic Press, New York, 1986.
5. R.D. McCarty, "Thermodynamic Properties of Helium-4 from 2 to 1500 K with Pressures up to 10^8 Pa", J. Phys. Chem. Ref. Data, 2(4): 923-1041, American Chemical Society, Washington, D.C., (1973).
6. R.B. Stewart, R.T. Jacobson and A.F. Myers, "The thermodynamic properties of Oxygen and Nitrogen", University of Idaho, Engineering Experiment Station, Final report on contract NAS 9-12078 (1972).
7. M. Ruhemann, "The separation of Gases", 2nd edition, Oxford University Press, New York, (1949).
8. M. Davies, "The Physical Principles of Gas Liquefaction and Low Temperature Rectification", Longmans, Green and Co. (1949).
9. Frank Pobell, "Matter and Methods at Low Temperatures", Springer Verlag, 1992.
10. F.J. Edeskuty and K.D. Williamson, "Liquid Cryogenics", Vol. 11, "Properties and Applications", CRC Press, Boca Raton, Fl, 1983.
11. J.G. Weisend II, "Handbook of Cryogenic Engineering", Taylor and Francis, Philadelphia, USA (1998).
12. NIST Standards Database 12, Thermophysical Properties of Pure Fluids – Version 3.1, National Institute of Standard and Technology, USA (1986).

3 CHAPTER

Closed Cycle Refrigerators and Cryocoolers

1. INTRODUCTION

In Chapter 2 we described the principles of liquefaction of gases to produce low temperatures. A second method of producing low temperatures is by using a closed cycle cooling system. When the cooling power is in the range of kilowatts the cooling system is referred to as a *Cryorefrigerator*. On the other hand, if the cooling power produced is below a few watts, the cooling system is generally called as a *Cryocooler*.

2. HELIUM REFRIGERATOR

The helium liquefier circuit described in Chapter 2 can be used with some simple modifications as a *Closed Cycle refrigerator*. Here the liquid helium after Joule Thomson expansion is circulated through the part to be cooled. The vaporized helium then passes through the cold side of the train of heat exchangers and returns back to the compressor. In large size charged particle accelerator systems as well as in thermonuclear fusion devices, big superconducting magnets are used. They need refrigeration power in kilowatts at 4.2 K or lower. Closed cycle helium refrigerators are then the only option available for cooling these devices.

A schematic of a 1 kW helium refrigerator system planned to be used in the Institute for Plasma Research at Ahmedabad for the Superconducting Tokomak project is described by Sarkar et al. in the Proceedings of the 18th International Cryogenic Engineering Conference held in 2000. This will be a helium refrigerator producing 650 W of refrigeration along with a liquefaction capacity of 200 l/hr. Larger helium refrigerators have been built and are operating successfully at various accelerator laboratories all over the world. The discussion on large-scale helium refrigeration systems will not be continued further since this is beyond the scope of this book.

3. CLOSED CYCLE CRYOCOOLERS

Like a heat engine, a cryocooler also makes use of a working substance, which undergoes a thermodynamic cycle of operations. However, the cooling cycle is in the reverse direction to that of a heat engine. In a *closed cycle cryocooler* the working substance absorbs a certain amount of heat, Q_c, from the sample at the cold temperature T_c and rejects a greater amount of heat, Q_W, at a warm temperature T_W. Since Q_W is larger than Q_c, external work W has to be done on the working substance in each cycle. The amount of work needed to extract unit quantity of heat at the cold temperature T_c, the ratio W/Q_c, is a measure of the effectiveness of the refrigerator.

The cryocooler based on Carnot cycle is the ideal refrigerator in which all processes are reversible. For this ideal *Carnot refrigerator,*

$$(W/Q_c)_{Carnot} = (T_W - T_c)/T_c \quad (3.3.1)$$

The performance of an actual cooling cycle is compared with that of the ideal Carnot cycle. The *coefficient of performance* (COP) of an actual cycle is defined as

$$\eta = (W/Q_c)_{Carnot} / (W/Q_c)_{actual} \quad (3.3.2)$$

and is expressed as a percentage. The higher the value of η the more closely the cycle approaches the ideal Carnot cycle.

A closed circuit cryocooler can be *multi-staged.* The principle of operation of a two-stage cryocooler is shown in Figure 3.1. In the second stage of the cooler, heat Q_c is absorbed at a temperature T_c and work W_2 is done on the working substance to raise its temperature from T_c to an intermediate temperature T_i.

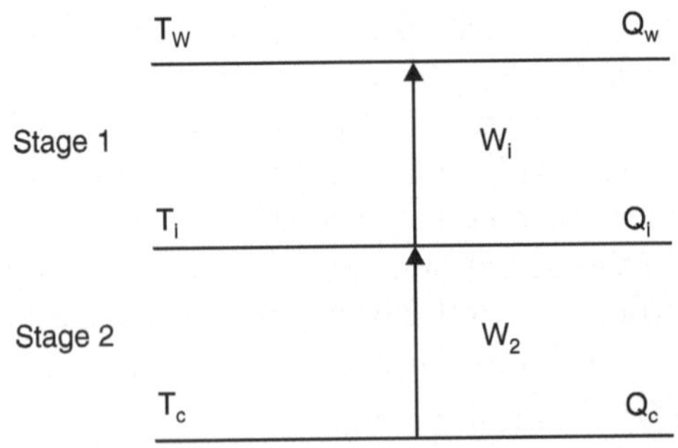

Figure 3.1: Multi-staged cryocooler system

This now acts as the cold end of the first stage of the cryocooler. At this intermediate temperature a quantity of heat Q_i may be absorbed from another object. The total energy in the working substance at T_i will be

$$Q = Q_c + W_2 + Q_i \quad (3.3.3)$$

Further work W_i is done on the working substance and heat Q_w is rejected at the warm temperature T_w.

The working substance in most cryocoolers is helium gas. In a single stage system, these cryocoolers can reach a temperature of around 50 K and in the two stage system, the second stage cold end (also known as cold head) can reach a temperature around 10 K. These are the temperatures obtained, when there is no heat load on the cold head of the cryocooler except for the minimal heat leak from the warm parts of the cryocooler. If an additional heat load Q is supplied at the cold head (say by activating a heater), the cold head temperature rises. The higher the applied heat load, the higher is the cold end temperature.

In the following we will use the words cryo-refrigerators and cryocoolers interchangeably.

There are many different thermodynamic cycles for a closed circuit refrigerator. We will describe here, only two most popular cycles, namely the Stirling and Gifford McMahon (GM) cycles. The cryorefrigerators built on these cycles have found world-wide acceptance. In both these cycles, the refrigerator consists of the following essential components.

(*i*) a compressor to compress the gas,

(*ii*) a displacer to move the gas from the cold to the warm end and vice-versa and

(*iii*) a regenerator, to enable the gas to exchange heat with the same.

But unlike the recuperative type heat exchanger, discussed in Chapter 2, this is regenerative type heat exchanger, which means the following. At any instant of time, there is only one stream of gas, either warm or cold, passing through the regenerator. The regenerator serves as a repository of heat from the warm gas when it passes through the regenerator and as a supplier of heat when the cold gas passes through it. This will be discussed further in a later part of this chapter.

The pulse tube cryocooler has been under development over the last two decades and has reached a stage where it can now compete with the other refrigerators for some applications. It differs from the above two cryorefrigerators, in that it has no displacer. A gas piston in the tube shuttles back and forth and serves the purpose of the displacer. The principle of operation of the pulse tube refrigerator will be briefly described in this chapter.

In contrast to the above refrigerators, the Joule-Thomson Cryocooler working down to 77 K usually operates in an open cycle. Although it needs a high pressure source of gas, it does not use a compressor. Its efficiency is poor compared to Stirling and GM cycles. It finds limited applications.

4. REGENERATOR

The regenerator should have the following characteristics. It should allow a relatively free passage for gas flow to ensure a low-pressure drop in the regenerator. The regenerator material must have a high thermal capacity in the temperature range in which it is used so that it can store heat efficiently. Also, it should have a sufficiently large thermal conductivity so that temperature uniformity in a direction perpendicular to fluid flow is achieved quickly. But there should be high resistance to heat flow in the direction in which the fluid is flowing. This can be achieved either by using spacers of low thermal conductivity between layers of the regenerator material or by the suitable design of the regenerator material itself.

For a single stage cryocooler operating down to ~30 K, the regenerator material will be a mesh of fine stainless steel or phosphor bronze wire. These metallic wires are 30 to 100 μm in diameter with mesh openings of the same size. Around 80 K, the thermal capacity of these materials is large and comparable to the classical value, 3 R per gram mole. But below this temperature the specific heat decreases rapidly.

For two stage cryorefrigerators, in which the second stage temperature is below 20 K, one should use a material with a lower Debye Temperature than copper. Lead is a suitable material. It is used in the form of small spherical balls (with diameters of the order of a few hundred μm).

Below 10 K, the thermal capacity of conventional regenerator materials is small compared to the thermal capacity of the working substance, namely, helium gas. This makes the regenerator efficiency quite low. This is the reason why the temperature range of operation of Stirling or GM refrigerators is usually limited to above 10 K.

Newer magnetic materials, which have a large magnetic specific heat at very low temperatures, have been developed for use as regenerator material in the second stage of the refrigerator. These materials order magnetically in the temperature range 4 to 15 K. Such an ordering is associated with a large peak in specific heat. Of these materials, *ErNi*, besides having a large enough specific heat anomaly, is chemically stable and also insensitive to oxidation. Using spheres of *ErNi* as regenerative material, it is possible to build two stage cryorefrigerators to provide refrigeration down to 4.2 K.

5. STIRLING CYCLE CRYOREFRIGERATOR

The Stirling cycle heat engine was first patented by Stirling brothers in 1816. The use of reversed Stirling cycle for cryorefrigeration was successfully developed in the Philips Research Laboratories in Eindhoven, Holland. In the Stirling cryogenerator, a piston and a displacer are mounted on the same shaft. The ideal cycle of operations consist of two isothermal and two isochoric (at constant volume) processes. The four phases of the reverse Stirling cycle are shown in Figure 3.2.

Phase 1: The displacer is at the top of the stroke and the piston is at the bottom of the stroke. The piston moves up while the displacer is stationary. The gas is compressed.

Phase 2: Piston is at the top of the stroke and the displacer is moving down. The gas is pushed from the warm space to the cold space first through a water cooler, in which the heat of compression is removed, and then a regenerator. The warm gas deposits its heat in the regenerator and enters the cold space at the top of the displacer.

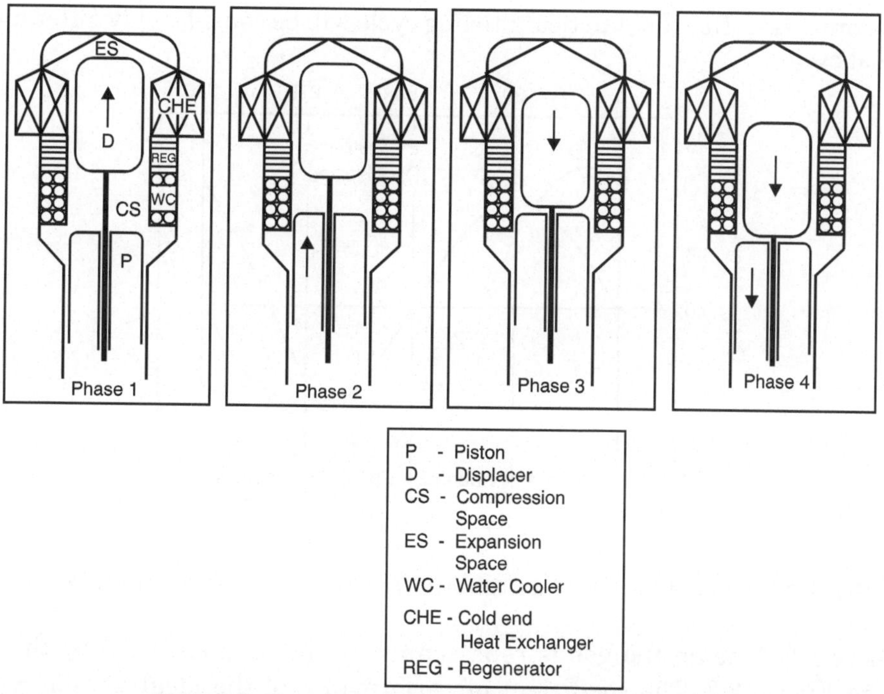

Figure 3.2: Four phases of the Stirling Cryorefrigerator cycle

Phase 3: After all the gas has been displaced to the cold space the piston and the displacer move down. This expands the gas in the cold space and produces cooling. The object to be cooled is attached to the cold head and exchanges heat with the cold gas.

Phase 4: The piston has reached the bottom of its stroke along with the displacer. Now, the piston is stationary and the displacer moves up. The cold gas is pushed through the regenerator to the warm space. The gas deposits its cold in the regenerator and enters the compression space.

The ideal reverse Stirling Cycle can be plotted on a PV diagram as shown in Figure 3.3. It can be seen that it consists of two isothermals and two constant

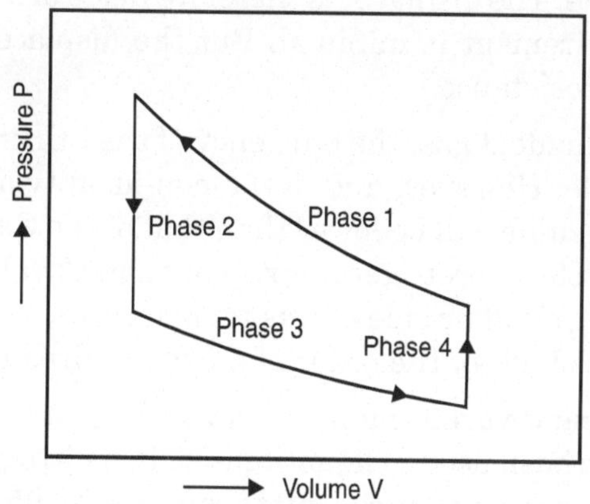

Figure 3.3: Ideal reverse Stirling cycle on a P-V Diagram

volume processes. But the practical Stirling cycle will be considerably different from the ideal cycle.

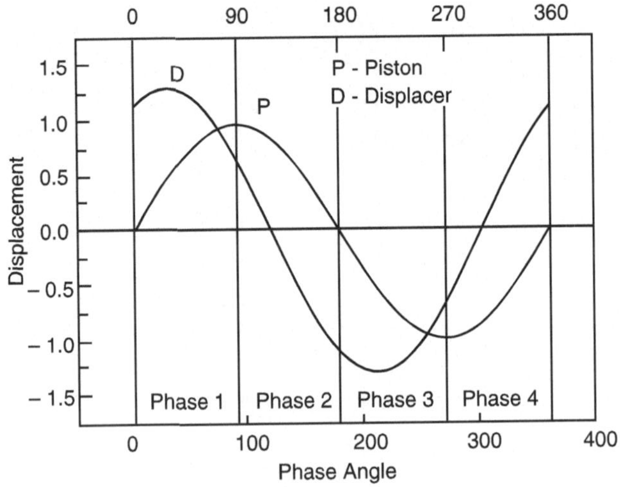

Figure 3.4: Motion of the piston P and displacer D in one cycle in the Stirling Cryogenerator

The work done on the gas is represented by the area enclosed by the closed curve in Figure 3.3. The coefficient of performance of the ideal Stirling Cycle is 100% i.e. it has the same efficiency as the Carnot cycle.

The discontinuous motion of the piston and displacer envisaged for the ideal cycle cannot be achieved in practice. The piston and displacer are driven by the same shaft. But their movements can differ in phase as shown in Figure 3.4. The four phases in the cycle are marked on the figure.

The piston has a large pressure difference across it. But the temperature difference across it is not large. So the piston must be sturdy and there should be a good fluid seal to prevent leakage of gas across it. The displacer, on the other hand, has a low pressure difference across it. But the temperature difference across the displacer is considerable. The displacer is therefore made of a light-weight material. The fluid sealing requirement is minimal. But the displacer should have a high longitudinal thermal resistance.

Theoretically, for an ideal gas, the efficiency of the Stirling cycle is the same as that of the Carnot cycle. However, due to the non-ideality of the cycle, the actual COP of the Stirling machine will be lower than that of the Carnot engine. The COP of a working Stirling cycle cryorefrigerator reaches a peak value of 40 to 50% around 125 K. The COP drops on either side of this temperature. The operating frequency of a Stirling machine is high-of the order of a few hundred to a few kiloHertz.

The integral Stirling cryogenerator, in which the piston and displacer operate in a single housing, has been used to liquefy gases. In a nitrogen liquefier a Pressure Swing Adsorption system produces nitrogen gas of 95% purity at ambient temperature. When this gas is let in at a pressure slightly above atmospheric pressure into the head of the cryogenerator, it gets cooled by contact with the cold head heat exchanger and gets liquefied. This liquid can be collected through a pipe through which it flows under gravity. Such liquefiers are now available in modular form. One can have a multiple cylinder cryogenerator to increase the capacity for liquefaction. Liquid nitrogen plants based on the Stirling cycle have been found to be rugged and reliable.

Figure 3.5: A photograph of a Stirling Cycle Nitrogen liquefier

5.1. Split Stirling Cycle Cryocooler

For certain applications like cooling IR detectors in space, one needs a refrigerating power of a few watts at about 77 K. But the system must be vibration free. Freedom from vibrations can be ensured by separating the piston from the displacer and connecting the compression space and the displacer space by a flexible coupling. Such an arrangement is called a split Stirling cycle cryocooler and is shown schematically in Figure 3.6.

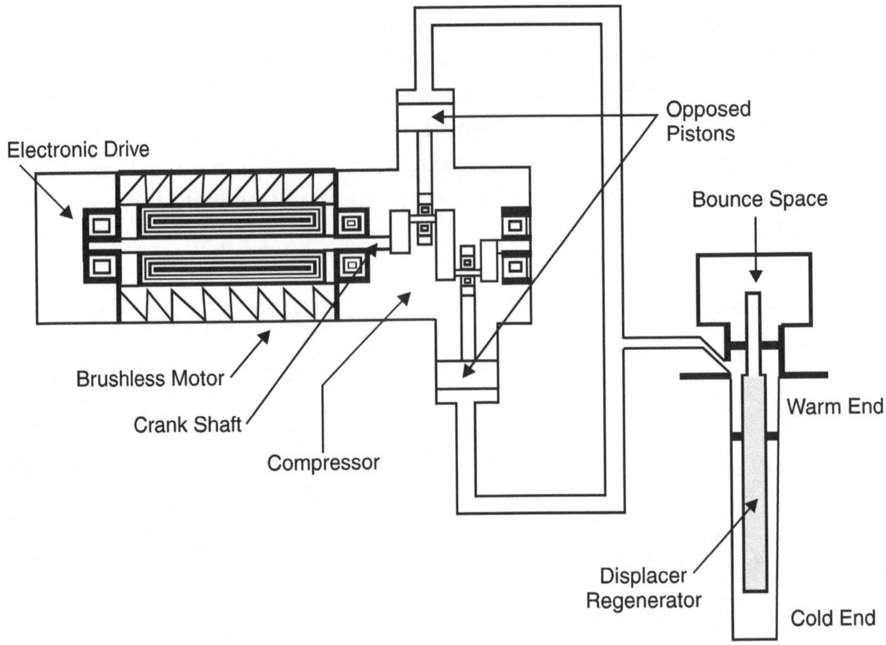

Figure 3.6: A split Stirling cycle cryocooler (from J.G. Weisand II, "Handbook of cryogenic engineering", Taylor and Francis, Philadelphia USA (1998)).

5.2. Free Displacer-Free Piston Stirling Cryogenerator

In the case of the integral Stirling cryogenerator, the piston and the displacer are mounted on a common shaft. There are kinematic links to convert the rotary motion of the flywheel into the linear motions of the piston and the displacer. Such a kinematic drive has bearings, seals and oil lubricant. This results in contamination of the working space with oil, and also friction and wear of the moving parts. If the moving components are rendered free of cumbersome kinematic linkages, the above problems can be mitigated. This is achieved in the free piston-free displacer (FPFD) type cryocoolers.

In the FPFD machines, the moving piston and displacer are driven linearly either pneumatically or electromagnetically. The moving part is suspended on a spring which determines its mean position. Because the driving force is axial, no side loading occurs. This eliminates the need for oil lubrication. Close tolerance dry seals can be used. This increases the life and reliability of FPFD cryocoolers.

In the electromagnetically driven compressors, the piston is driven by a linear motor. The piston is attached to a coil placed in a transverse magnetic field. When an *AC* current is passed through the coil, the coil moves back and forth linearly and drives the piston. The piston is suspended on a spring. The free displacer moves through the differential pressure of the gas. The phase difference in the motion of the displacer relative to the compressor piston is adjusted by springs and mass loading.

The main areas of application of such Stirling coolers are in cooling (*i*) *IR* detectors, and (*ii*) filters of high temperature superconductors for mobile telephony. Cryocoolers which produce ~4 to 6 W at 80 K are now commercially available and have a mean life time between failure longer than 20,000 hours.

6. GIFFORD-McMAHON REFRIGERATOR

A schematic of the GM refrigerator is shown in Figure 3.7. Helium gas is compressed in the compressor *C* to a pressure of about 120 psi. The compressor is the usual domestic refrigeration compressor modified to remove the larger heat of compression of helium.

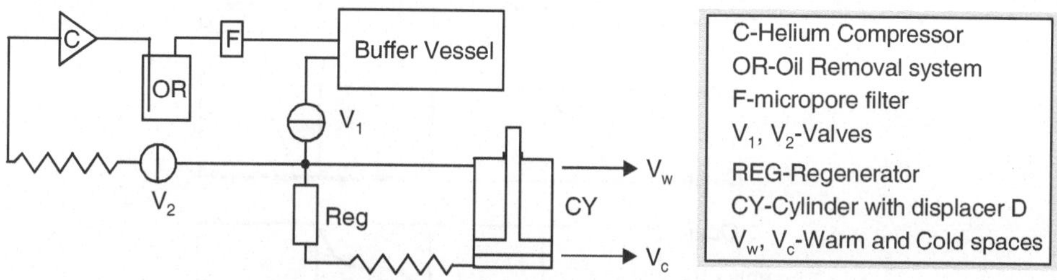

Figure 3.7: Schematic diagram of a GM Refrigerator

Oil is injected in the compressor suction line to remove the heat of compression. After removing the heat of compression, oil vapour in the helium gas is removed through three successive separation devices, namely centrifugation, coalescer with a return line at low pressure for oil droplets and a micropore filter *F*. Traces of oil vapour in the helium will freeze in the colder parts of the refrigerator and block the flow of gas. The micropore filter plays a very important role in removing the final traces of oil vapour in the gas. The compressed gas flows into a buffer vessel.

The cycle consists of the following four phases. The variation of pressure, position of the displacer and valve openings V_1 and V_2 in the four phases are shown in the figure 3.8.

Phase I: The displacer *D* in the cylinder *CY* is at the bottom of its stroke. Valve V_2 is closed and valve V_1 is opened. The pressure rises from P_2 to P_0. The temperature of the gas in the volume V_W above the displacer rises above 300 K due to adiabatic compression.

Phase II: The valve V_1 remains open, the displacer moves up. The gas in the space V_W moves through the regenerator *R* to the cold space V_C. Since the temperature of V_C is lower than the temperature of V_W additional gas is drawn from the buffer vessel through the regenerator. This process is isobaric at constant pressure P_0.

Phase III: Valve V_1 closes quickly while valve V_2 opens at the same time. The displacer remains in the top position. The pressure drops.

Phase IV: The displacer moves down to its original position. The cold gas is pushed through the regenerator to the compressor through the cold station where heat is extracted from the sample. This process is not exactly isobaric due to heat exchange in the regenerator.

The compressed helium gas is connected to the cold finger through flexible pressure lines. The displacer in the cold finger is either mechanically or

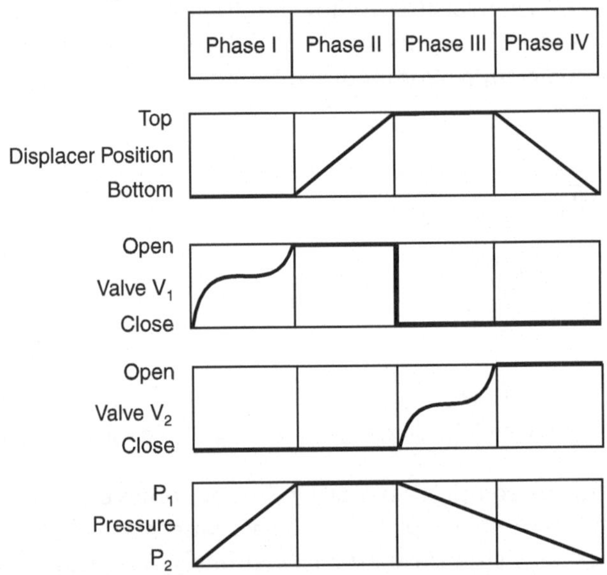

Figure 3.8: Variation of displacer position, valve openings and pressure in a GM cycle

pneumatically driven. The synchronisation of the motion of the displacer with the opening and closing of the *HP* and *LP* valves is achieved by using a rotating distributor.

The advantages of the GM refrigerator are the following:

- The compression and expansion volumes are separate. This results in reduced vibrations of cold parts.
- The operating frequency is of the order of a few hertz. Because of this low frequency operation wear and tear of the components is less.
- The displacer operates with very little pressure difference on either side of it. There can be more tolerance in the clearance of the displacer.
- Multi-staging the GM refrigerator is simple.

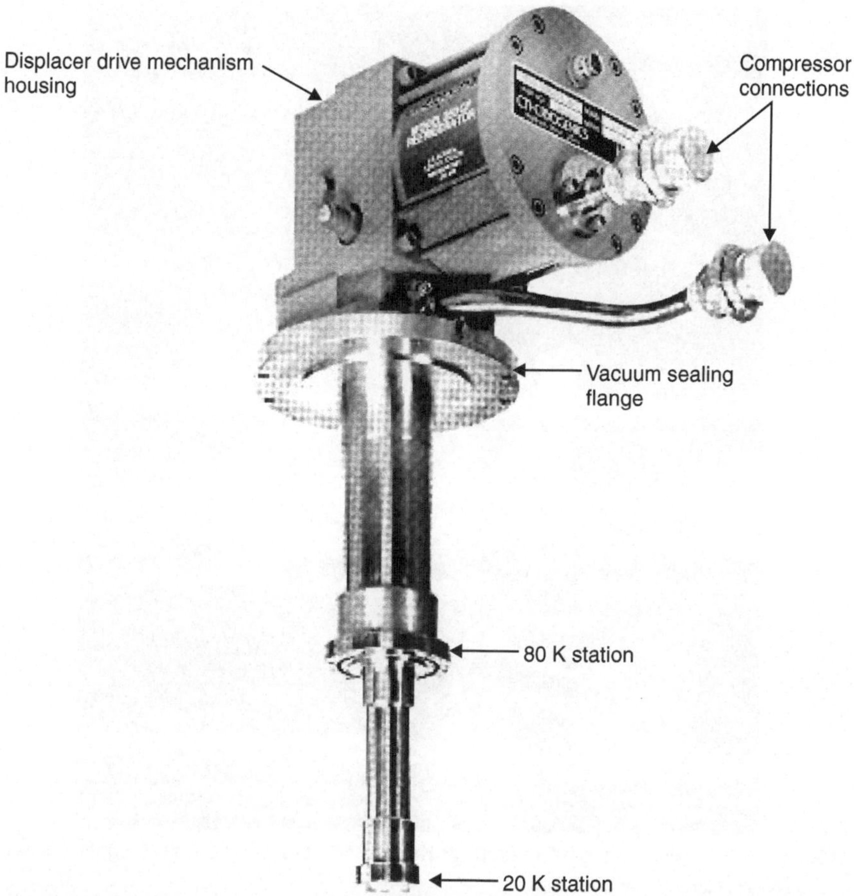

Figure 3.9: Photograph of a two stage commercial GM refrigerator, manufactured by CTI Cryogenics, USA. It produces 3 W at 20 K combined with 10 W at 85 K in the first stage. The refrigerator has a length of 470 mm and weight of about 10 kg. Helium compressor is not shown in the above figure.

The areas of application of GM refrigerators are in cryopumping, cooling of radiation shields in MRI magnet cryostats, and cooling samples for carrying out experiments in the laboratory.

Two stage GM refrigerators producing about 20 W refrigeration around 100 K and 1 W refrigeration around 10 K are commercially available (See Figure 3.9) and are being used for low temperature measurements in many laboratories in the country.

More recently GM refrigerators have been used to cool superconducting magnets. Conventional superconducting magnets need to be dipped in a liquid helium bath for cooling. In the dry HTSC magnet systems the need for such a bath is eliminated. Figure 3.10 shows a picture of such a cryogen-free superconducting magnet.

Figure 3.10: Conduction-cooled magnet offered by Kobe Steel, Ltd., and Japan Magnet Technology, Inc., www.kobelco.co.jp/indexe.htm.

7. PULSE TUBE REFRIGERATOR

High reliability in small cryocoolers has been a problem which has been studied for many years. One approach to increase the reliability is to eliminate some of the moving parts in a mechanical refrigerator. The Stirling type refrigerator has two moving components, namely, the piston and the displacer, where as the GM type refrigerator has the displacer as the moving component.

In 1964, Gifford and Longsworth [8] discovered a new refrigeration technique, which eliminated the moving components of the Stirling system by replacing it with the gas piston. This is known as Pulse Tube Refrigeration (PTR). Work on PTR has advanced significantly over the last two decades. As on date, commercial systems are available, which produce refrigeration powers similar to those of GM type systems for several applications.

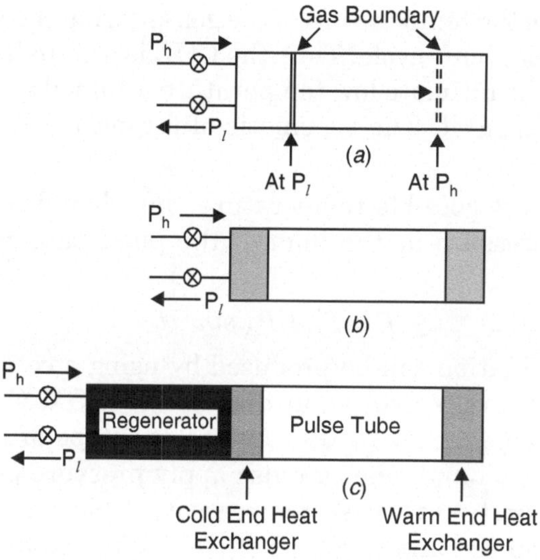

(*a*) Empty tube closed at one end with pressure pulses applied, (*b*) The same arrangement with heat exchangers at their ends, (*c*) The above arrangement with regenerator before the entry of pressure pulses.

Figure 3.11: Working principle of a Pulse Tube Refrigerator

The working of a PTR can be understood with the help of Figure 3.11. Consider a thin walled cylinder closed at one end and high and low pressure being alternately applied at the other end. The gas boundary at low pressure will move from the left end to the right end of the tube as the pressure inside the tube increases from P_l to P_h. The work of compression is now transferred to the gas and hence the gas will get heated when it moves to the right. Since the gas is in contact with the wall of the tube, the closed end of the tube gets heated when the gas is compressed. Similarly, as the pressure decreases from P_h to P_l, the gas boundary returns back to the original position. When the pressure decreases from P_h to P_l the gas is cooled and hence the wall of the tube in contact with the gas at the open end is also cooled. Repeated compression and expansion of the gas or pressure pulsing causes heating and cooling of the gas to occur alternately and this in turn causes an asymmetrical effect, namely, the heating to occur at the closed end and the cooling to occur at the open end.

When the heat exchangers are added at either end as shown in Figure 3.11 (*b*), the effect is now substantiated, since warm end heat exchanger (WHE) ensures that the gas is maintained at some steady temperature after compression. Hence, on expansion, the cold end heat exchanger (CHE) shows some cooling effect.

To convert the above observed effect into useful refrigeration, a highly efficient regenerator is introduced as shown in Figure 3.11 (*c*). The WHE is now circulated with cooling water. By this, the heat of compression at the end of high pressure P_h will be removed from WHE, causing the gas to come back to ambient temperature. Now the gas will get cooled on expansion and the gas will gradually cool both CHE and regenerator. The next batch of incoming gas takes in the cold stored in the regenerator and hence the temperature of the gas entering the Pulse Tube is slightly less than that of the previous cycle. Thus the CHE is gradually cooled to lower and lower temperature. The ultimate low temperature attained at the cold end depends on the heat load entering it. The schematic of a typical PTR is shown in Figure 3.12.

To attain the lowest possible temperature, one should use helium gas as the working fluid. The pressure at the inlet of the pulse tube refrigerator is varied sinusoidally as

$$P = P_O + P_A \sin \omega t \qquad (3.7.1)$$

This pressure oscillation can be produced by using a compressor and a rotary distribution valve (Rotary Valve) as in the case of a GM Cryocooler. When this frequency of oscillation ω is low (within 10 Hz), this type of PTR is known as GM type PTR. On the other hand, one can also apply pressure oscillations directly to the Pulse Tube system by valveless compressors. This occurs at higher circular frequency (~ within 50 Hz) and this type of PTR is known as Stirling type PTR.

The pressure wave produces a sinusoidal oscillation in the velocity υ of the gas in the pulse tube. Let the velocity be given by

$$\upsilon = \upsilon_0 \sin(\omega t - \phi) \qquad (3.7.2)$$

The enthalpy flow at any point in the pulse tube averaged over one period of the pressure cycle is given by,

$$< H > = \frac{C_p}{\tau}\int_0^{\tau} \dot{m}\, T dt = \left(\frac{C_p A}{R}\right)(1/\tau)\int_0^{\tau} \upsilon P\, dt \qquad (3.7.3)$$

Here the mass flow rate $\dot{m}$ is $\varrho A\upsilon$, where ϱ is the gas density, A is the area of cross section of Pulse Tube and υ is the velocity of flow. Using ideal gas equation, ϱ is replaced as P/RT. Herein τ refers to the period of oscillation of the pressure wave and C_p is the specific heat of the gas in the tube. The average enthalpy flow is proportional to Cos (ϕ) where ϕ is the phase difference between the pressure and velocity fields. The average enthalpy flow is a maximum when the pressure and velocity fields at any point are in phase. Thus, the theory based on enthalpy balance indicates that *if heat needs to be extracted from the cold end of the Pulse Tube, then the velocity and pressure oscillations must be in phase at this end.*

The basic Pulse Tube system of Gifford and Longsworth, wherein the valves V_1 and V_2 of Figure 3.12 are closed, did not meet the above criterion and hence the refrigeration achieved was miniscule. In fact, the phase difference is nearly $\pi/2$. The cooling produced in this case is essentially due to gas to wall conduction heat transfer. One can invoke an analogy between the velocity and the pressure in the pulse tube to the electric current and voltage in an electrical circuit. In the electrical circuit, the phase difference between the voltage and current can be controlled by a proper choice of resistance, capacitance and inductance.

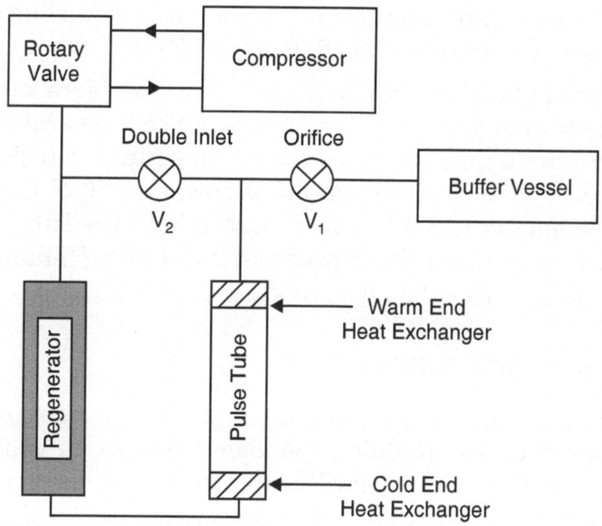

Figure 3.12: Schematic of a typical Pulse Tube Refrigerator

In the pulse tube refrigerator, the role of resistance is played by the orifice needle valve V_1 and the role of the capacitance by the buffer vessel. By adjusting the opening of the orifice and the volume of the buffer vessel one can adjust the phase difference ϕ to approach zero in GM type pulse tube coolers operating at low frequency. Mikulin [9] in 1984 proposed the concept of the *Orifice type PTR*. Alternately, by introducing a capillary between V_2 and the buffer volume, one can also control the phase ϕ. This capillary, called the inertance tube, is equivalent to an inductance in an electrical circuit. This method of control is used in Stirling type pulse tube refrigerators operating at high frequency.

Zhu et al., [10] in 1990 proposed the *double inlet* concept. Here one introduces the flow impedance V_2 to allow a small flow of gas at the warm end of the pulse tube, thereby reducing the gas flow through the regenerator. This improved the performance of the orifice type Pulse Tube refrigerators. One was able to reach lower temperatures.

However, the double inlet pulse tube also causes a parasitic dc flow through the regenerator and the pulse tube and this reduces its ultimate performance as a cryocooler. This dc flow can be cancelled by connecting the orifice side to the low-pressure side of the compressor through a needle valve. This flow tends to cancel the parasitic dc flow through the pulse tube. This improves the performance of the double inlet pulse tube refrigerator. In Stirling type Pulse Tubes, it is more difficult to cancel the dc flow through the pulse tube. In such cases, double inlet arrangement is not used and it is sufficient to control the phase difference by the orifice mode itself.

The pulse tube refrigerator has the advantage that there is no moving displacer in the tube and this considerably reduces the vibrations of the cold tip. However the COP of the PT refrigerator is low compared to that of a Stirling or GM cryocooler. Efforts are underway to improve the efficiency of PTR.

For space borne applications, Stirling mode PT cryocoolers have been developed to produce a refrigerating power of a few watts in the temperature range between 50 to 80 K. For ground applications, GM type Pulse tube refrigerators have been developed providing ~100 W of refrigeration power at 80 K and nearly 10 W at 15K. With suitable magnetic regenerator materials Pulse Tube Refrigerators can now produce ~0.5 W of refrigeration power at 4.2 K. Using Helium 3 as the working fluid, a temperature of 1.78 K has been obtained with PTRs.

8. JOULE-THOMSON CRYOCOOLER

It was pointed out in Chapter 2, that in large liquefiers/refrigerators, the last stage of cooling is achieved by isenthalpic expansion through an orifice. One can also build open cycle cryocoolers, notably for IR detector cooling, using the Joule Thomson effect. Here the cooler is integrated with a detector dewar.

High pressure nitrogen gas (about 150 bars) enters the orifice and expands. The cold gas passes through a heat exchanger surrounding the inlet tube connected to the orifice. This provides cooling of the incoming gas. After a short time, the gas liquefies in the dewar. The JT orifice and heat exchanger shown in Figure 3.13 are mounted in a double walled evacuated dewar.

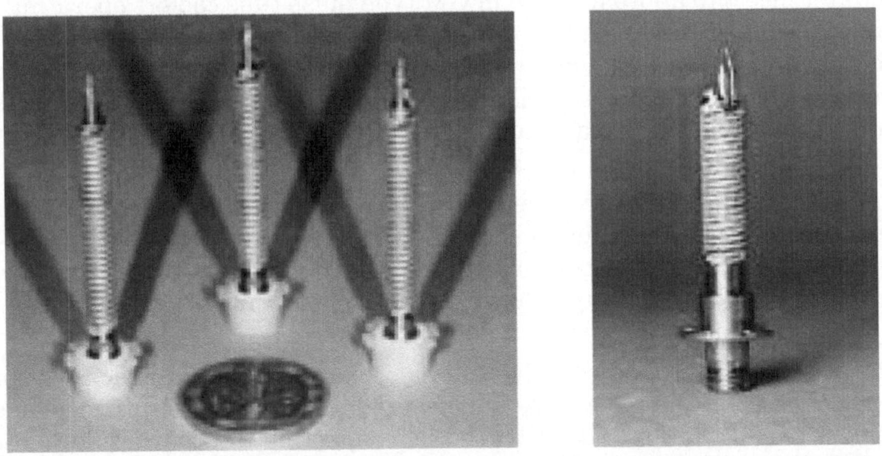

Figure 3.13: An integral Joule-Thomson Cryocooler for IR detectors (*Courtesy:* Catalogue of M/s Thales Cryogenics, France, www.thales-cryogenics.com)

The COP of such an open cycle JT cooler is poor compared to other closed cycle cryocoolers. However the advantages are the reliability of their operation, the stability of the temperature and the capacity for miniaturisation. Alfeev proposed in 1973 that using mixtures of gases one could reduce the pressure to which the mixture has to be compressed to achieve the same refrigeration capacity. For example, a mixture of nitrogen with hydrocarbons needs to be compressed only to

90 bar. This proposal has been verified. The JT cooler efficiency has been considerably improved. A new generation of more efficient JT coolers using gas mixtures is in the offing.

For a JT cooler operating at liquid helium temperatures the incoming gas must be precooled below 50 K by a two stage GM refrigerator. The gas flowing through the JT valve can be drawn from the GM compressor at its discharge stage. The expanded gas will have to be compressed to the suction pressure of the GM compressor before entry into it.

9. RELATIVE EFFICIENCIES OF DIFFERENT CRYOCOOLERS

Figure 3.14 compares the COPs of different types of cryocoolers as a function of temperature. It can be seen that the isothermal expansion of the Carnot's cycle accounts for its 100% efficiency. The adiabatic expansion with work recovery has nearly 80% efficiency in most of the temperature regions except below about 20 K. On the other hand, adiabatic expansion without work recovery has relatively lower efficiency and gradually increases with lowering of temperature upto about 20 K.

The thermodynamic efficiency of orifice Pulse Tubes fall in the 40 to 70% range, especially above 75 K. The JT expansion systems have lower efficiencies compared to other systems. The efficiency of the JT system for a given fluid increases with temperatures approaching its liquefaction. One can use different fluids such as N_2, H_2 and Helium for different temperature ranges.

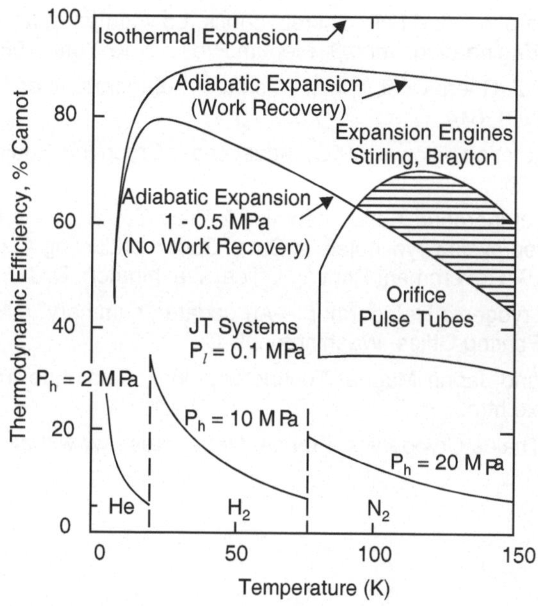

Figure 3.14: Relative efficiencies of different cryocoolers as a function of temperature (from Ray Radebaugh, "A review of pulse tube refrigerators", Advances in Cryogenic Engineering **35**, (1998), 1191-1205).

10. CONCLUSION

Small closed cycle refrigerators are used quite commonly in many laboratories. The advantage with these refrigerators is that a desired low temperature can be reached by the push of a button. The need for relatively large liquefiers with problems of maintaining and servicing them is eliminated. The development of these closed cycle refrigerators must be considered revolutionary as they have made low temperature measurements accessible economically to small laboratories, especially in the underdeveloped countries.

REFERENCES

1. B.A. Hands, "Cryogenic Engineering", Academic Press, New York, 1986.
2. Randall F. Barron, "Cryogenic Systems", 2nd edition, Oxford University Press, New York, 1985.
3. Thomas M. Flynn, "Cryogenic Engineering", Marcel Dekker, Inc., New York, 1997.
4. G. Walker and E.R. Berigham, "Low capacity cryogenic Refrigeration", Clarendon Press, Oxford, 1994.
5. G. Walker, "Cryocoolers", Vol. I & II, Plenum Press, New York, 1983.
6. G. Walker, "Stirling cycle Machines", Oxford University Press, New York, 1973.
7. G. Walker, "Miniature refrigerators for cryogenic sensors and cold electronics", Monographs on Cryogenics, Oxford University Press, New York, 1989.
8. W.E. Gifford and R.C. Langsworth, (1964). Pulse tube refrigeration progress, "International Advances in Cryogenic Engineering", Vol. 10, Pt. 2, Plenum Press, New York.
9. E.I. Mikulin, A.A. Tarasov, and M.P. Shkrebyonock, Low temperature expansion pulse tubes, Adv. in Cryogenic Engineering, vol. 29, Plenum Press, New York, 1984, pp. 629-637.
10. S. Zhu, P. Wu, and Z. Chen, Double inlet pulse tube refrigerators: an important improvement, *Cryogenics* 30, (1990), 514.
11. W.H. Kohler, Refrigeration below –100°C, "Advances in Cryogenic Engineering", Vol. 5, Plenum Press, New York, 1960.
12. R. Radebaugh, "Refrigeration Fundamentals: A View Toward New Refrigeration Systems, Application of Closed-cycle Cryocoolers to Small Superconducting Applications", NBS special publication 508, U.S. Government Printing Office, Washington, D.C.
13. T.R. Strobridge, "Cryogenic Refrigerators—An updated summary", NBS Technical note 655, U.S. Government Printing Office, Washington, D.C.
14. Kobe Steel, Ltd., and Japan Magnet Technology, Inc., Catalogue from Net, address www.kobelco.co.jp/indexe.htm.
15. Catalogue of M/s Thales Cryogenics, France, Net address www.thalescryogenics.com.

CHAPTER 4

Mechanical Properties of Solids

1. INTRODUCTION

The choice of materials for construction of cryogenic equipment is dictated by consideration of their mechanical and thermal properties at these temperatures. Since the properties at low temperatures are significantly different from those at room temperature, there is no substitute for the actual data. Fortunately there are several data compilations now available (McClintock and Gibons (1960), Durham et al (1962), Campell (1974) and Johnson (1960)). The data on mechanical properties can be better appreciated, when one understands the effects of the application of stresses on solid structures.

2. STRENGTH AND DUCTILITY

When a bar of a solid is subjected to a tensile or elongating force, there will be an elongation proportional to the applied force as long as the force is small. The material will very nearly return to its original length, when the force is released. This is called elastic behavior of the solid. All solids exhibit this in some measure. At some higher values of stress, the material will not behave elastically. It may break without further deformation (this is known as *brittle behavior*), or take on a permanent deformation (this is known as ductile or *plastic behavior*).

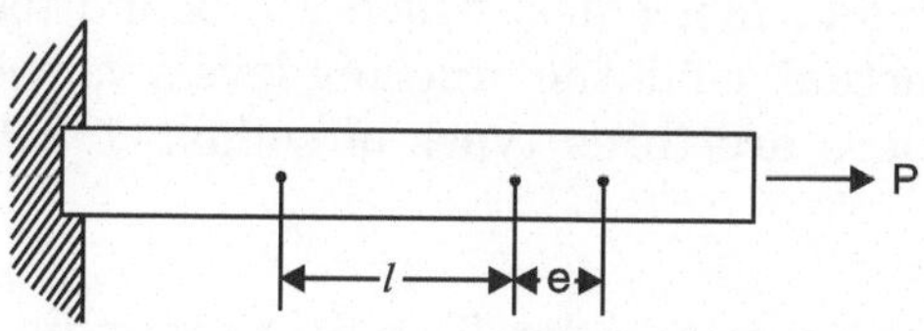

Figure 4.1: Stress-strain test on a specimen

The familiar stress-strain test is shown in Figure 4.1, in which a rod of gauge length l is subjected to a force P along its length. The stress is defined as the force per unit area of cross section of the rod, *i.e.*

$$S = P/A \tag{4.2.1}$$

where A is the area of cross section.

Due to this load, the gauge length l extends by an amount e. The tensile strain ε is defined as,

$$\varepsilon = e/l \tag{4.2.2}$$

The relationship between S and ε is known as Stress-strain curve. Such curves are shown in Figures 4.2(*a*) and (*b*) for a ductile and brittle material. The linear part of the curve up to *PL* corresponds to the elastic region and the rest of the curve corresponds to the plastic region of the material.

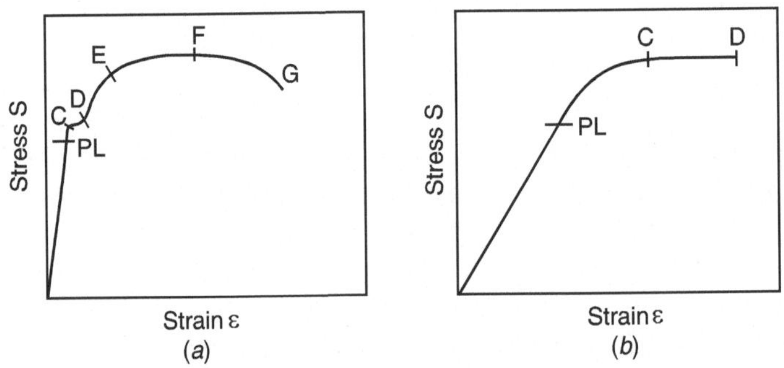

Figure 4.2: The Stress-strain relationship of (*a*) Ductile and (*b*) Brittle materials

Another class of materials (non-metals) that are capable of extreme elastic behavior also exists and they are called Elastomeric materials. In actual practice, materials, which exhibit plastic deformation, are more desirable for structures. This is due to the fact that if the stress gets increased accidentally beyond the design value, it will be redistributed by plastic flow of the material. On the other hand, brittle materials have no such mechanism to protect the structures from excessive stresses.

To understand the mechanism causing ductility in solids and its dependence at low temperature, the *structure* of solids needs to be understood. The theory of elasticity treats the solid as a continuous medium although it is not really so. The solid comprises of atoms or molecules, which are bound together in more or less regular arrays. In an actual solid, the ordering levels vary depending on the type of solid. Essentially there are three types of solids. (1) Glasses (2) Plastics and (3) Metals.

Glasses: Glasses lack the spatial order. To a first approximation, the ordering in a glass is similar to that of a viscous liquid with short range ordering.

Plastics: Plastics are polymers composed of long chain molecules. Each chain can have hundreds to ten thousands of atoms. The intermolecular forces that bind the polymer molecules are very weak and they are the van der Waals forces. The cross-linking between the chains is by the formation of chemical bonds. When the level of cross linkages is low, the plastics are known as thermoplastics. When this level is high, they are known as thermosetting plastics.

Metals: Metals exhibit highly ordered structure. They are arranged in symmetrical crystal lattices. The three common crystal lattice structures encountered with metals are Face Centered Cubic (FCC), Body Centered Cubic (BCC) and Hexagonal Closed Packing (HCP). They are shown in Figure 4.3. The number of slip planes is highest in FCC and is lowest in BCC.

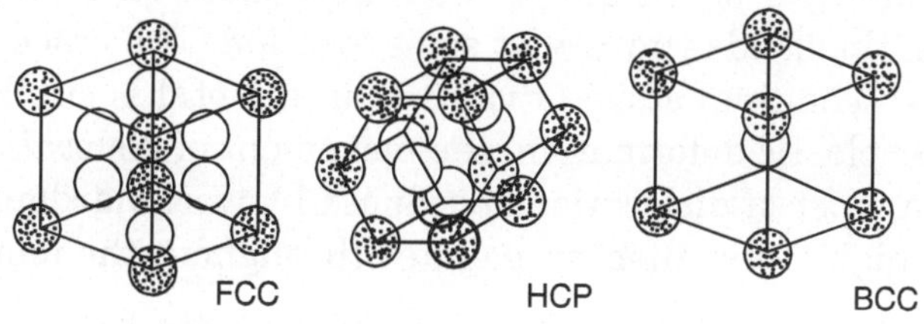

Figure 4.3: Three most common crystal structures

In real crystals a perfect arrangement of atoms as shown in Figure 4.3 does not exist. Irregularity in structure can arise due to different factors such as impurity atoms, vacancies (where atoms are missing), and additional interstitial atoms. These are examples of point defects (PD) in the crystal. Extended imperfections known as dislocations may be present in a crystal. Dislocations are of two types. *Edge dislocations* arise when there is an extra plane of atoms extending part of the way in the crystal. *Screw dislocations* arise when a part of the crystal has slipped one atomic distance relative to its adjacent part.

The concept of the geometrically perfect crystal is still useful in explaining some properties of the crystal; but there are many other properties that can be adequately explained only by invoking defects. For instance, the concept of dislocations helps to explain the Yield Stress and ductility of metals, whereas the perfect crystalline structure helps to explain the Modulus of Elasticity.

One can account for elasticity by imagining a regular crystal structure. When a horizontal shearing stress is applied at the top of the crystal, the structure will shear; but it will return back to its original shape, when the above stress is removed. However, beyond a certain stress value, plastic shear will take place along one of the planes of atoms resulting in a permanent deformation of the structure.

If the stress values required to shear a perfect crystal are calculated theoretically on the basis of inter-atomic forces, these values are very much higher than the experimentally measured values. However, the experimental values match very well with those required to move a dislocation through a crystal lattice.

The ability of crystalline materials to deform plastically (*i.e.* the ductility of the material) depends primarily on the mobility of dislocations in a crystal, which, in turn, depends upon the temperature. The number of slip systems available in a crystal also influences the dislocation mobility. (The slip systems refer to the directions in which the planes can slip over one another).

Impurity atoms, vacancies and interstitial atoms will impede the movement of the dislocations. The smaller the grain size of a cluster of atoms, the higher is the yield strength of the material. The reason for annealed metals to be more ductile than the work-hardened metals is that work hardening reduces the size of the grains.

Glasses have no slip systems due to their amorphous structure. When stress is applied to a glass, the bonds rupture at some location. The crack then propagates causing a fracture. The final stage of rupture in a metal is similar to that in the case of glass. After plastic deformation, the movement of dislocations is blocked by tangles of large number of dislocations produced by work hardening. The rupture at the location of high stress then propagates throughout the material as a crack.

3. DUCTILITY AT LOW TEMPERATURE

The effect of temperature on ductility arises, to a first approximation, from the effect of temperature on the slip planes and the motion of dislocations. Let us first consider the case of metals. The ductility of FCC metals is relatively insensitive to decrease of temperature. The metals of other crystal structure tend to become brittle at low temperature. This is because FCC structure has more slip systems than other crystal structures. Also, the slip planes of BCC and HCP crystals tend to change at low temperature. Hence copper, copper-nickel alloys, aluminum and its alloys, and austenitic stainless steels containing more than 7% nickel, which have the FCC structure, remain ductile down to low temperatures.

Iron, carbon and low alloy steels, molybdenum and niobium having the BCC structure become brittle at low temperatures. The HCP-based systems occupy an intermediate place between the above two.

Plastics and elastomers do not respond to stress as metals do. The less cross-linked elastomers yield by uncoiling their long chain molecules, which can slide over one another. This is facilitated by the thermal energy at room temperature.

At low temperatures, the attractive forces are more effective than the thermal energy and hence the material deformation is considerably restrained. This effect is pronounced by its "glass transition" at the onset of brittleness. In view of the above, all elastomers become brittle at low temperatures with the exception of polytetrafluroethylene (PTFE), which deforms plastically down to 4 K.

3.1. Variation of Ductility with Temperature

The typical variation of ductility (*i.e.* the elongation in a tensile set) versus temperature is shown in Figure 4.4. In general an S shaped transition between two temperatures T_1 and T_2 is observed.

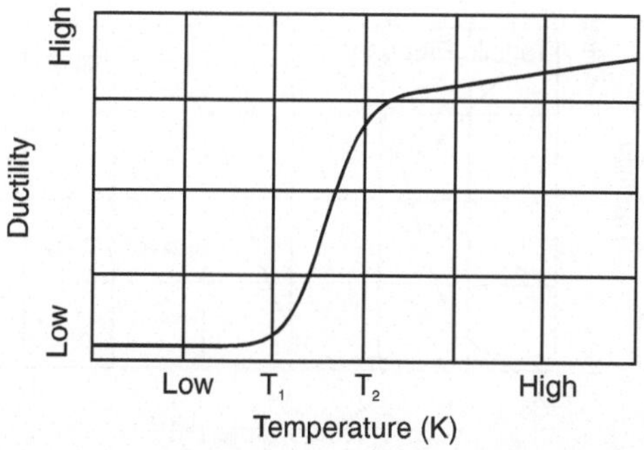

Figure 4.4: The general curve of ductility with temperature

The transition region is the range of temperature, where the response mechanism to stress is gradually becoming inactive. The percentage elongation before failure of several structural materials is shown in Figure 4.5. Of the various metals, stainless steel shows increased ductility with increasing temperature. On the other hand, carbon steel indicates a transition at about 100 K. This is the ductile to brittle transition occurring in this material at this temperature range.

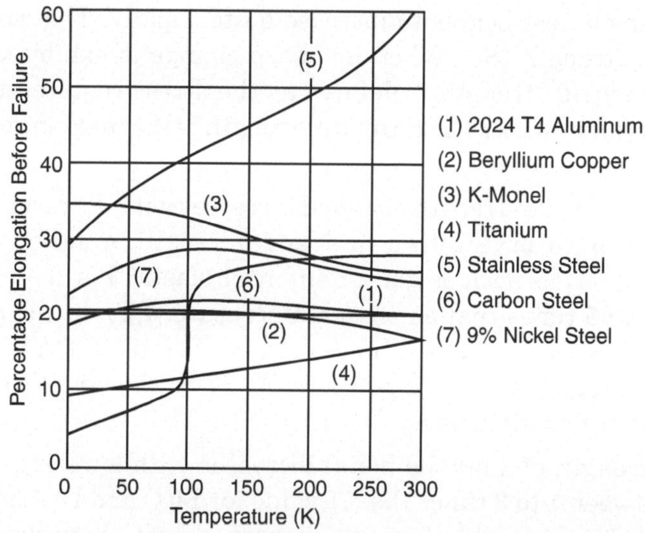

Figure 4.5: The curve of ductility versus temperature for several important structural materials. (1) 2024 T4 Aluminum, (2) Beryllium copper, (3) K-Monel, (4) Titanium, (5) Stainless Steel, (6) Carbon steel and (7) 9% Nickel steel (from Thomas M. Flynn "Cryogenic Engineering", Marcel Dekker, Inc., New York, (1997)).

Tensile and impact tests are generally used for evaluating the effect of strain rate and stress complexity on the ductility behavior of materials. The tensile test applies a unidirectional stress at a slow rate, while the impact test applies a multidirectional stress at a rapid rate. The ductile behavior varies for impact and tensile stresses at a given temperature, as can be seen in Figure 4.6.

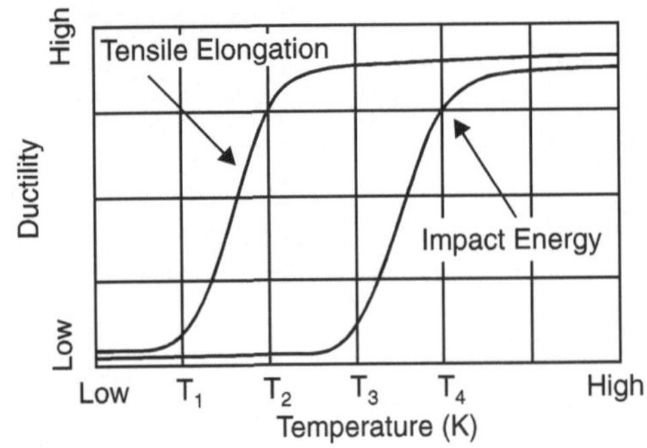

Figure 4.6: Effect of strain rate and stress complexity (from Thomas M. Flynn "Cryogenic Engineering", Marcel Dekker, Inc., New York, (1997)).

4. LOW TEMPERATURE STRENGTH OF SOLIDS

4.1. Ultimate and Yield Strengths

For many materials, there is a definite value of stress at which the strain of the material in a tensile test begins to increase quite rapidly. This value of stress is defined as yield strength (S_y). When no sharp change is exhibited in the stress-strain curve, the yield strength is defined as the stress required to permanently deform the material by 0.2%. The ultimate strength is the maximum nominal stress attained during a tensile test.

In general, the tensile strength of a solid increases with decreasing temperature. This is true for both crystalline and non-crystalline solids and for many heterogeneous materials such as glass-reinforced plastic. For metals, the strength at 4 K will be 2 to 5 times that at room temperature (RT). In the case of plastics, the strength at 77 K will be 1.5 to 8 times greater than the RT value. Glasses show less change in strength at low temperature. At 77 K, the strength is only 1.5 to 2 times that at room temperature.

The yield strength of a ductile metal increases with lowering of temperature. At 20 K, it is between 1 to 3 times the RT value for FCC and HCP metals. For BCC structured metals, the yield strength increases quite rapidly on lowering of temperature and it becomes greater than the fracture strength, leading to fracture of the material before the yield strength is reached.

The yield and ultimate strengths of copper are shown in Figure 4.7. This shows that the yield strength can be 'insensitive' to "work hardening" of the metal. However, in the case of carbon steel shown in Figure 4.8, the yield and tensile strength approach each other on lowering the temperature. The material becomes brittle at the temperature where the two curves converge.

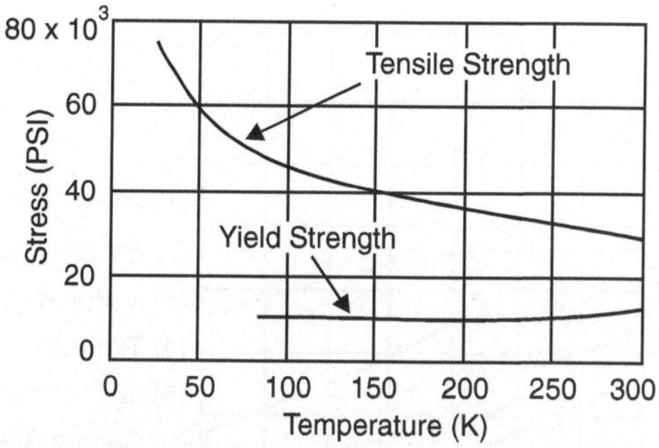

Figure 4.7: Tensile and yield strengths of annealed oxygen free copper (from Thomas M. Flynn "Cryogenic Engineering", Marcel Dekker, Inc., New York, (1997)).

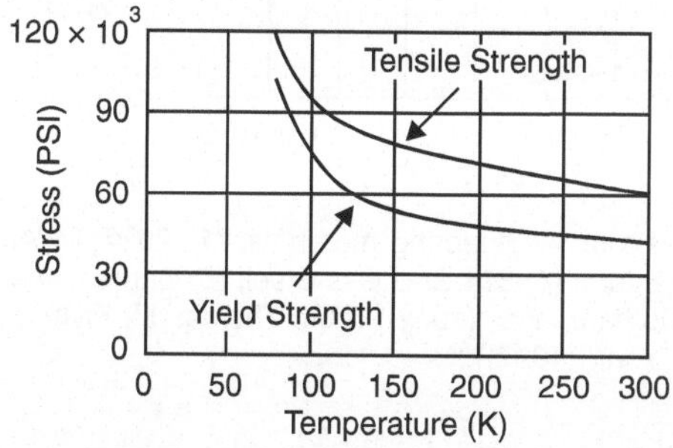

Figure 4.8: Tensile and yield strengths of plain carbon steel (SAE1010m), cold roll (from Thomas M. Flynn "Cryogenic Engineering", Marcel Dekker, Inc., New York, (1997)).

Many engineering materials are alloys with alloying atoms having a different size compared to that of parent material. When the alloying atoms are smaller in size, they tend to migrate and pin down the dislocations. This causes an increase in yield strength.

On lowering of temperature, the atoms vibrate less vigorously and hence a larger stress needs to be applied to move the dislocation and this causes an increase in the yield strength at low temperature. The temperature variation of yield strength of several engineering materials is shown in Figure 4.9. The behavior of ultimate strength is also similar and hence not presented here.

Only a few plastics have been tested below 200 K. As mentioned earlier, Teflon shows ductility down to 4 K. However, reinforced plastics such as glass fiber laminates exhibit good properties, such as increased tensile strength in a direction parallel to the lamination, at lower temperature.

Figure 4.10 shows the tensile strength of several plastics. Increase in strength is accompanied by a rapid decrease in elongation and impact resistance. The glass-reinforced epoxies are quite useful for cryogenic applications due to their high strength to weight ratios.

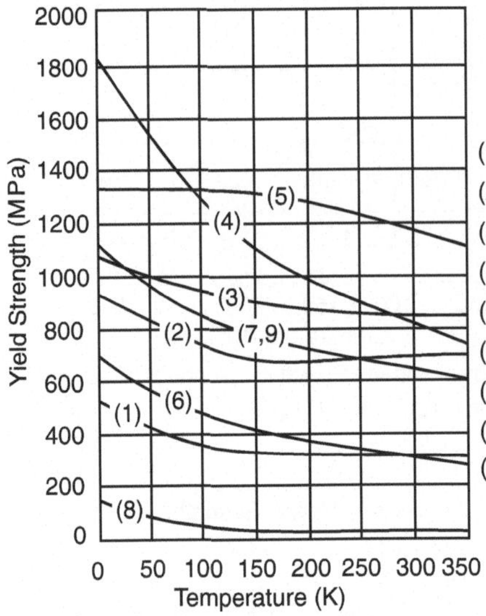

Figure 4.9: Yield strength of several engineering materials, (1) 2024 T4 Aluminum, (2) Beryllium copper, (3) K-Monel, (4) Titanium, (5) 304 Stainless Steel, (6) Carbon steel, (7) 9% Nickel steel, (8) Teflon and (9) Invar 36 (Durham et al. (1962)) (from Thomas M. Flynn "Cryogenic Engineering", Marcel Dekker, Inc., New York, (1997)).

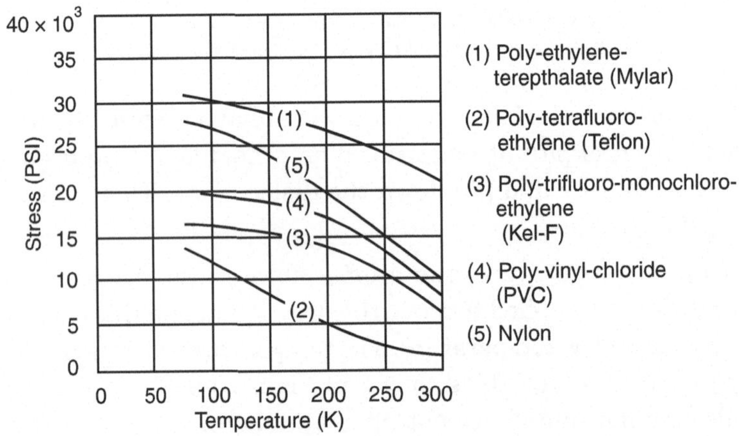

Figure 4.10: Tensile strength of selected plastics. (from Thomas Flynn "Cryogenic Engineering", Marcel Dekker Inc., New York, (1997)).

4.2. Fatigue Strength

The stress required for failure after a given number of cycles is called the fatigue strength (S_f). A simple reversed bending test is the usual method of measuring fatigue strength. For some materials, such as carbon steel fatigue failure will not occur, if the stress is kept below a certain value, called endurance limit S_e, no matter how many reverse cycles are carried out. Figure 4.11 shows the fatigue strength of several engineering materials at 10^6 cycles.

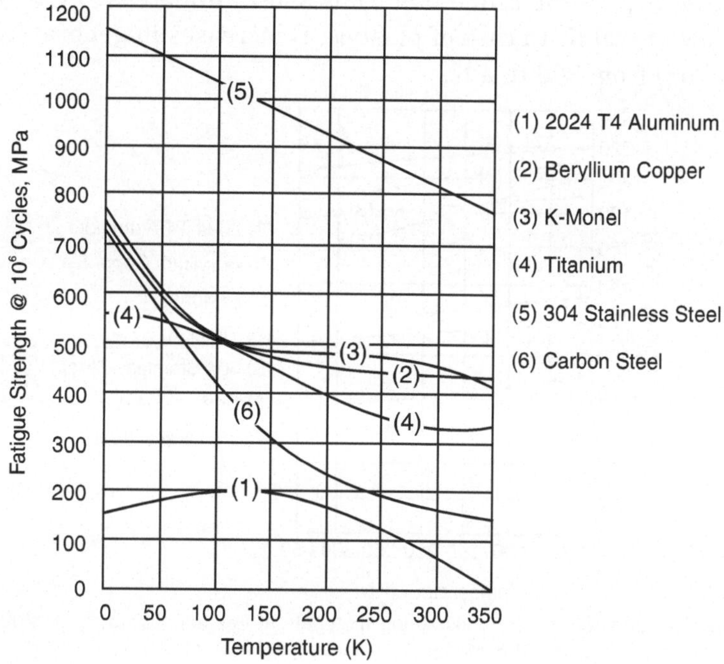

Figure 4.11: Fatigue strength at 10^6 cycles for several engineering materials, (1) 2024 T4 Aluminum, (2) Beryllium copper, (3) K-Monel, (4) Titanium, (5) 304 Stainless Steel, (6) Carbon steel and (Durham (1962)).

5. ELASTIC MODULI

There are basically three commonly used elastic moduli. They are

- Young's modulus (E), which is the ratio of tensile stress with respect to tensile strain at constant temperature in the elastic region,
- Shear modulus (G), which is the rate of change of shear stress with respect to shear strain at constant temperature in the elastic region and
- Bulk modulus (B), which is the rate of change of pressure corresponding to the change in volume per unit volume at constant temperature.

When the material is isotropic, the moduli of elasticity are related by Poisson's ratio ν, the ratio of strain in the direction of the uniaxial stress to the strain in the perpendicular direction.

$$B = E/[3(1 - 2\nu)] \tag{4.5.1}$$

$$G = E/[2(1 + \nu)] \tag{4.5.2}$$

The variation of Young's modulus with temperature of several materials is shown in Figure 4.12. It is observed that the elastic modulus increases with decreasing temperature. Most polycrystalline materials show an increase in E by about 10% from 300 K down to 20 K. In case of plastics, E increases in general from 2 to 20% for *PTFE* (Teflon) from 300 to 4 K.

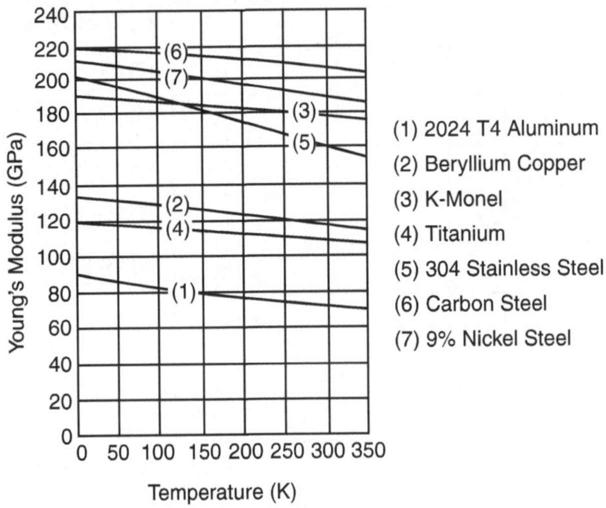

Figure 4.12: Young's Moduli of several engineering materials at low temperatures, (1) 2024 T4 Aluminum, (2) Beryllium copper, (3) K-Monel, (4) Titanium, (5) 304 Stainless Steel, (6) Carbon steel and (7) 9% Nickel steel (Durham (1962)).

6. CONCLUSION

It is convenient to classify metals by their lattice structure for low temperature mechanical properties. The FCC structure metals and their alloys are most often used in the construction of cryogenic equipment. Aluminum, Copper, Nickel, and their alloys and austenitic stainless steels of the 18-8 type are FCC and do not exhibit a ductile to brittle transition at low temperatures. As a general rule, the structural properties of these metals improve at lower temperatures. The yield strength, Young's modulus and fatigue strengths are also improved at lower temperatures (with the exception of 2024 T4 aluminum).

The BCC structured metals and alloys are usually undesirable for low temperature construction. This includes iron, martensitic steels (Carbon and 400 series stainless steels), Molybdenum and Niobium. They exhibit a ductile to brittle transition at low temperatures.

HCP structured metals exhibit structural properties in between those of FCC and BCC metals. Titanium and its alloys with HCP structure remain reasonably ductile down to low temperature.

Plastics also increase in strength as the temperature is decreased. But this is accompanied by a rapid decrease in elongation in tensile test and decrease in impact resistance as the temperature is lowered. The glass-reinforced plastics have high strength to weight and strength to conductivity ratios. All elastomers become brittle at low temperatures. Many materials including rubber, Mylar and Nylon can be used as static seal gaskets provided they are highly compressed at room temperature prior to cooling.

The strength of glass under constant loading also increases with decrease in temperatures. Glass with surface defects fails at lower stress values, but can be improved by tempering the surface.

REFERENCES

1. H.G. Van Bueren, "Imperfections in Crystals", Interscience Publishers, New York, 1960.
2. H.P. Gibbson and R.M. McClintock, "Mechanical properties of materials at low temperatures", NBS Monograph 13, US Government Printing Office, Washington D.C.
3. Thomas M. Flynn, "Cryogenic Engineering", Marcel Dekker, Inc., New York, 1997.
4. Randall F. Barron, "Cryogenic Systems", 2nd edition, Oxford University Press, New York, 1985.
5. B.A. Hands, "Cryogenic Engineering", Academic Press, New York, 1986.
6. K.A. Warren and R.P. Reed, "Tensile and Impact properties of Selected Materials From 20 to 300° K", NBS Monograph 63 (1963).
7. R.P. Reed and A.F. Clark (eds), "Materials at Low Temperatures", ASM, Ohio (1983).
8. D.A. Wigley, "Mechanical Properties of Materials at Low Temperature", Plenum Press (1971).
9. T.F. Durhan, R.M. McClintock and R.P. Reed, "Cryogenic Materials Data Hand book", Office of Technical Services, Washington D.C.
10. McClintock, M., "Cryogenics", Reinhold Publishing Corp., New York, 1964.

5

CHAPTER

Thermal Properties of Materials at Low Temperatures

1. INTRODUCTION

The mechanical properties of materials were discussed in Chapter 4. In this chapter we shall discuss the thermal properties of materials.

2. THERMAL PROPERTIES OF MATERIALS

The most important thermal properties are (*a*) specific heat, (*b*) thermal expansion and (*c*) thermal conductivity. The thermal capacity of a material determines the amount of cryogen required to cool a given mass of material to a specific low temperature. Thermal expansion is important especially when joints of dissimilar materials are made. When the joint is cooled, differential contraction of dissimilar materials can lead to thermal stresses, which might cause a mechanical failure of the joint. Thermal conductivity determines how rapidly heat is transported from one part of the material to another. A rapid dissemination of heat in a material leads to a quick equalization of temperature and this will be an advantage in some cases. However since one end of a cryostat will be at room temperature, thermal conduction leads to heat flow from the warm end to the cold end. This heat flow should be as small as possible in order to reduce the consumption of cryogen.

3. SPECIFIC HEAT

At any temperature the atoms in a solid are vibrating about their equilibrium positions. If there are N atoms in a solid these vibrations can be resolved into 3 N independent normal modes. Each normal mode has a frequency ν. At a temperature T the energy of the normal mode at the frequency ν is given by

$$\varepsilon(\nu) = h\nu/(\exp(h\nu/k_B T) - 1) \tag{5.3.1}$$

The frequencies of the normal modes for a solid will be closely spaced. We may define a frequency distribution function $g(\nu)$ such that the number of normal modes between ν and $\nu + d\nu$ is $3\,Ng(\nu)d\nu$. The frequency distribution function is normalized *i.e.*

$$\int_0^{\nu_m} g(\nu)d\nu = 1 \tag{5.3.2}$$

ν_m is the maximum frequency of the vibrational modes.

The thermal energy of the solids is

$$U_{\text{th}} = 3N \int_0^{\nu_m} [h\nu/(\exp(h\nu/k_B T) - 1)]\, g(\nu) d\nu \qquad (5.3.3)$$

The *specific heat at constant volume* of the material is given by

$$C_v = (\partial U_{\text{th}}/\partial T)_V = 3Nk_B \int_0^{x_m} \{x^2 e^x/(e^x - 1)^2\}\, g(x)\, dx \qquad (5.3.4)$$

where

$$x = h\nu/k_B T \qquad (5.3.5)$$

The specific heat of any material decreases as the temperature is decreased. If the material can be treated as an elastic continuum, then the specific heat of the material can be written as

$$c_v = 3nk_B D(\Theta_D/T) \qquad (5.3.6)$$

Here n is the number of atoms in one gram of the material and T is the absolute temperature. *The Debye temperature,* Θ_D, of a material characterizes the temperature dependence of specific heat. If we take a material containing an Avogadro's number N of atoms then, n is now replaced by N and hence, the specific heat C_v is given as,

$$C_v = 3R\, D(\Theta_D/T) \qquad (5.3.7)$$

For a monatomic solid, C_v corresponds to the molar specific heat. The function $D(x)$ is called the *Debye specific heat function* and is defined as

$$D(x) = (3/x^3)\int_0^{x} [y^4 e^y/(e^y - 1)^2]\, dy \qquad (5.3.8)$$

The plot of the Debye curve is shown in Figure 5.1. In this graph C_v is plotted as a function of T/Θ_D. When T is less than $\Theta_D/20$, $D(\Theta_D/T)$ varies as $(T/\Theta_D)^3$. The specific heat of any material varies as T^3 at very low temperatures.

Table 5.1 shows the Debye temperatures of a few common materials. Note that diamond, which is the hardest material known, has the highest Debye temperature. Similarly graphite has a higher Debye temperature than copper. At low temperature the specific heats of diamond and sapphire are much lower than the specific heat of copper. Lead, which is a soft material, has a low Debye temperature. Its specific heat at low temperature is much higher than that of copper. It is the reason why lead is used as a regenerator material in cryocoolers operating down to 10 K.

The contribution to specific heat from lattice vibrations is called the phonon contribution. In insulators only phonons contribute to the thermal energy. On the

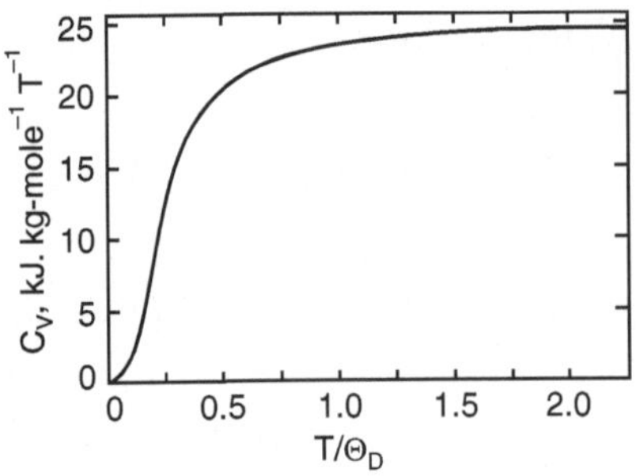

Figure 5.1: Temperature dependence of Lattice Specific heat C_V according to Debye Theory. Here, Θ_D is the Debye Characteristic Temperature

other hand in metals both the free electrons and phonons contribute to the thermal energy. The *electronic contribution*, $C_{v\,el}$, to the total specific heat C_v is a term proportional to the temperature and can be written as,

$$C_{v\,el} = \gamma T \qquad (5.3.9)$$

This term should be added to the phonon contribution. At room temperature the phonon contribution dominates, since the electronic contribution is only a few percent of the phonon contribution. But the phonon contribution, which varies as T^3 at very low temperatures, decreases rapidly with falling temperature and becomes less than the electronic contribution at temperatures below 1 K. The value of γ for metals is also given in Table 5.1.

Table 5.1: Debye Temperatures of different materials

The coefficient of T in the electronic specific heat of metals is also given in the Table. γ is in units of mJ/mole/K^2.

Material	*Debye Temp. (K)*	γ	*Material*	*Debye Temp. (K)*	γ
Diamond	2200	—	Gold	165	0.729
Graphite	1500	—	Silver	225	0.646
Copper	345	0.695	Lead	96	2.98
Stainless Steel	320	4.98	Indium	111	1.69

So far the discussion has been on the specific heat C_v at constant volume. Experimentally one measures the specific heat, C_p at constant pressure. The following thermodynamic relation connects C_p and C_v.

$$C_p - C_v = TV\beta^2/\chi_T \qquad (5.3.10)$$

Here V is the molar volume of the material, β is the volume expansion coefficient and χ_T is the isothermal compressibility of the material. The difference, $C_p - C_v$, will amount to a few percent of the total specific heat at room temperature. The difference decreases as the temperature falls and becomes negligible at very low temperatures.

The specific heat of a material is important for low temperature applications in two ways. These are discussed below. To cool a material of mass M from room temperature, 300 K to a lower temperature, T, we have to remove from the material a quantity of heat Q, given by

$$Q = \int_T^{300} Mc_p dT \tag{5.3.11}$$

This heat can be removed by using the latent heat of vaporisation of the liquid at its boiling point or by using the total enthalpy of the cryogen from boiling point to 300 K.

Usually the sample in a cryogenic experiment is under high vacuum. Still there is a heat leak from the surroundings to the sample. If the refrigeration provided by the cryogen is cut off, the sample will warm up due to this heat leak. It is usual to assume the heat leak to be given by $K\Delta T$, where $\Delta T(t)$ is the difference in sample temperature $T_s(t)$ from the surroundings temperature T_0 (i.e. $[T_0 - T_s(t)]$) and K is an empirical constant.

The sample temperature will rise, when the refrigeration is cut off and is given by,

$$T_s(t) = T_0 - \Delta T_{(t=0)} \exp(-t/\tau) \tag{5.3.12}$$

The relaxation time τ, which is the time in which the difference in temperature falls to $1/e$ of its initial value, is given by Mc_p/K. The relaxation time is directly proportional to the thermal capacity of the sample. At low temperature the thermal capacity becomes small and the sample will warm up at a fast rate. The relaxation time will increase as the temperature is increased.

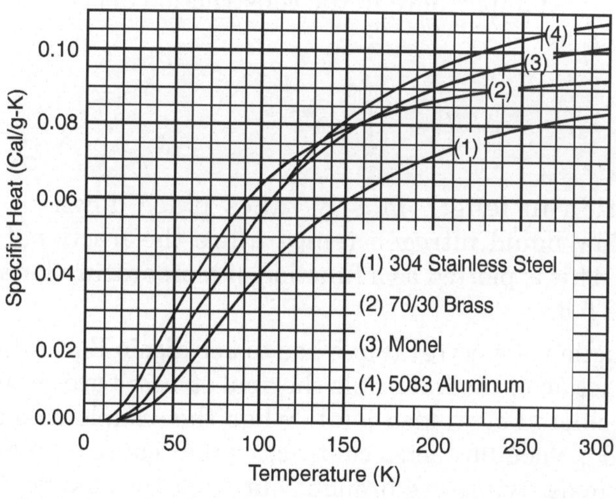

Figure 5.2: Specific heat as a function of temperature for different materials (from I. Klipping (6))

In some situations, the cooling process is a single shot process as in adiabatic demagnetization. One would like to perform an experiment on the sample after it is cooled to a temperature T by such a process. The time in which the experiment is done should be much shorter than the relaxation time if the temperature of the sample should not change appreciably during the experiment. One should take stringent precautions to reduce K, i.e., to reduce heat leak into the system.

Figure 5.2 depicts the specific heat as a function of temperature for several materials.

4. THERMAL EXPANSION

Usually a material contracts in volume as the temperature is reduced. The volume expansion coefficient of a material is defined as

$$\beta = (1/V)(\partial V/\partial T)_p \tag{5.4.1}$$

For an isotropic material the volume expansion coefficient β is three times the linear expansion coefficient α defined as

$$\alpha = (1/L)(\partial L/\partial T)_p \tag{5.4.2}$$

The linear expansion coefficient of a material is a function of temperature. It decreases as the temperature decreases in a way roughly similar to the behaviour of the specific heat. The Gruneisen law states that the ratio G given by,

$$G = \beta V/\chi\, C_v \tag{5.4.3}$$

for a material is a constant independent of temperature. While this law is obeyed at high temperature, deviations from this law are observed at low temperatures.

The fractional contraction in length, $\Delta L/L$, as the material is cooled from 300 K to a temperature T is of importance in the construction of cryostats. This fractional contraction is given by

$$\Delta L/L = -\int_T^{300} \alpha\, dT \tag{5.4.4}$$

The contraction in length mostly takes place down to liquid nitrogen temperature. Below liquid nitrogen temperature the fractional contraction of a material is small. This is plotted as a function of temperature for a few materials of interest in Figure 5.3.

In the construction of a cryostat one has to use dissimilar materials. Due to the differential contraction of the materials thermal stresses are set up. This may lead to a failure of a soldered or welded joint. When the sample chamber in a cryostat has to be operated in vacuum such a contraction may cause a failure of the vacuum. Sometimes thin copper wires are brought out from the sample, which is at a low temperature, to the top flange, which is at room temperature, to make electrical measurements. If enough slack is not given to these wires, they may snap when cooled, leading to a loss in electrical connectivity.

It is also noteworthy that polymer materials like teflon or nylon have a contraction which is roughly an order of magnitude more than that of a metal. When a polymer material is used along with a metal in the construction of a cryostat the design must clearly allow for the large difference in contraction between the two materials. This point is discussed in Chapter 6.

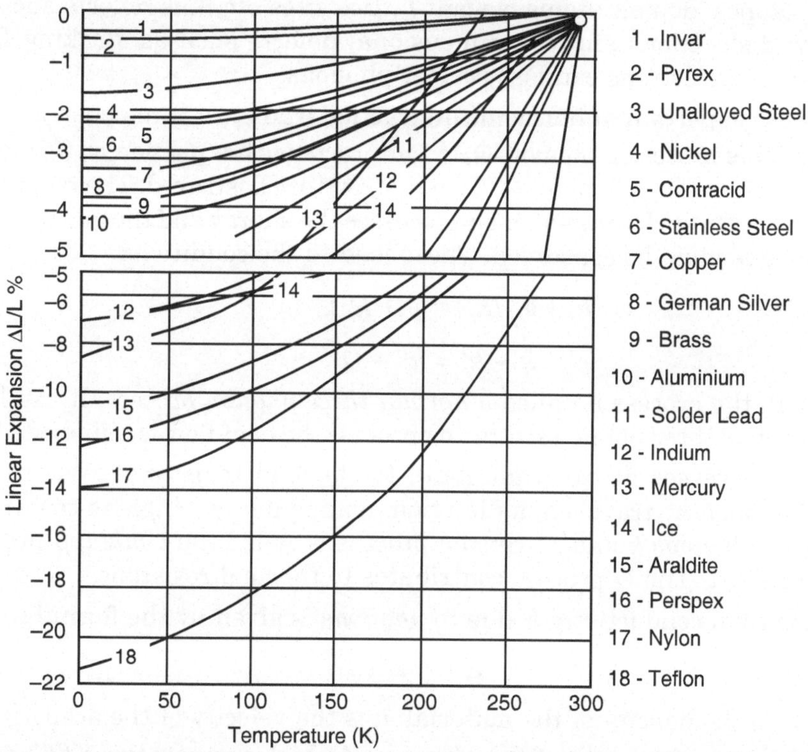

Figure 5.3: Contraction in length $\Delta L/L$ from 300 K to a temperature T is plotted as a function of T for several materials (from I. Klipping (6)).

If a cryostat works satisfactorily down to liquid nitrogen temperature one may assume that differential contraction will not cause any problem if the cryostat is cooled to lower temperatures like 4.2 K.

5. THERMAL CONDUCTIVITY

The heat flux Q in watts/m^2 is proportional to the temperature gradient, grad T, in a material.

$$Q = -K \text{ grad } T \tag{5.5.1}$$

The constant of proportionality K is called the thermal conductivity. It is expressed in W/mK. In a solid heat is conducted by phonons. Just as photons are quanta of electromagnetic radiation, phonons are quanta of lattice modes of vibration. A phonon, corresponding to a normal mode of frequency $\omega_j(\boldsymbol{k})$, carries an energy $(h/2\pi)\,\omega_j(\boldsymbol{k})$ and a momentum $(h/2\pi)\boldsymbol{k}$. The wave vector, $\boldsymbol{k}$, is restricted by convention to the first Brillouin Zone. A normal mode of wave vector $\boldsymbol{k}$ is identical to a mode of wave vector $\boldsymbol{k} + \boldsymbol{G}$, where $\boldsymbol{G}$ is a reciprocal lattice vector of the lattice. Phonons can be scattered due to the anharmonic interactions of the lattice and due to defects in the lattice. These defects include zero dimensional defects (point defects) such as isotopes, dopant atoms, vacant lattice sites etc, line defects such as edge and screw dislocations and two-dimensional defects such as stacking faults. In metals the free electrons can also scatter phonons.

The most important anharmonic interaction leads to a three phonon scattering process. In this process a phonon $\omega_j(\boldsymbol{k})$ interacts with a phonon $\omega_{j'}(\boldsymbol{k}')$ to produce a phonon $\omega_{j''}(\boldsymbol{k}'')$. The phonons $\omega_j(\boldsymbol{k})$ and $\omega_{j'}(\boldsymbol{k}')$ are destroyed. The reverse process is also possible. In such three-phonon processes the energy and the quasi-momentum of the phonons must be conserved. These lead to the conditions

$$\omega_j(\boldsymbol{k}) + \omega_{j'}(\boldsymbol{k}') = \omega_{j''}(\boldsymbol{k}'') \tag{5.5.2}$$

and

$$\boldsymbol{k} + \boldsymbol{k}' = \boldsymbol{k}'' + \boldsymbol{G} \tag{5.5.3}$$

If $\boldsymbol{G} = 0$, the process is called a *normal three phonon process*. It can be shown that in a normal three phonon scattering process the heat flow is unimpeded. Normal scattering processes do not contribute to thermal resistance. When $\boldsymbol{G} \neq 0$ the scattered phonon $\boldsymbol{k}''$ travels in a direction making a large angle to the direction of the incident phonons $\boldsymbol{k}$ and $\boldsymbol{k}'$. Such a process is called the *Umklapp* process or U process, in short. The U process contributes to thermal resistance.

The thermal conductivity K due to *phonons* is given by the formula

$$K = \rho\, c_v\, v\, l \tag{5.5.4}$$

Here ρ is the density of the material, v is the velocity of the acoustic phonons and l is the mean free path of the phonons. At high temperatures ($T >> \Theta_D$), the mean free path is determined by the number of phonons which are available to scatter the incident phonon. This number is proportional to k_BT. The mean free path varies as $1/T$. At high temperatures c_v tends to a constant value. The density ρ and the velocity v are weakly temperature dependent. So the thermal conductivity varies as $1/T$ at high temperature.

The $\boldsymbol{k}$ vector of one of the phonons involved in scattering must be more than half way to the edge of the Brillouin zone for U process to occur. This implies that at least one of the phonons must be a phonon of frequency larger than $\omega_D/2$. The number density of such phonons will decrease exponentially as $\exp(-(h/2\pi)\,\omega_D/2k_BT)$ or $\exp(-\Theta_D/2T)$. The mean free path therefore increases as

$\exp(\Theta_D/2T)$ as the temperature T falls. At the same time the specific heat c_v decreases as the temperature is reduced. So the thermal conductivity will vary as

$$K(T) \propto T^a \exp(\theta_D/bT) \tag{5.5.5}$$

The exponent b has a value close to 2 and the exponent, a, depends on temperature.

We have so far considered only the U process to be responsible in determining the mean free path l. There will be other scattering processes due to defects. Point defects, like Frenkel and Schottky defects, are always present in a material in thermal equilibrium. Point defects scatter phonons, and the scattering cross section varies as the fourth power of the frequency. One and two-dimensional defects also scatter phonons. Their scattering cross sections will have different dependences on the frequency of the phonons.

When all scattering processes are present, the effective mean free path l can be written as

$$1/l = 1/l_U + \Sigma_j 1/l_j \tag{5.5.6}$$

l_U is the mean free path due to three phonon U-processes and l_j is the mean free path due to the j-th type of defect in the crystal. While l_U is intrinsic to the crystal and increases rapidly as the temperature is reduced, l_j is inversely proportional to the number density of the j-th type of defect and shows a less rapid dependence on temperature. The effective mean free path is determined by the smaller of the values l_U and l_j. As one lowers the temperature from 300 K, the thermal conductivity will show a maximum value, K_m, at a temperature T_m. On reducing the temperature further, K will decrease from the value K_m and tend to zero value as the temperature tends to zero. The peak temperature T_m will be determined by the temperature at which l_U becomes comparable to l_j. The value of K_m will be determined by the value of l_j at T_m. The larger the density of defects, the higher is T_m and the lower is the value of K_m.

At very low temperatures l_j may exceed the width of the sample. When this happens the phonon mean free path will depend on boundary scattering and will be equal to the width of the sample. The specific heat c_v will vary as T^3. So the thermal conductivity K will also vary as T^3. The value of K will depend on the size of the sample and nature of its boundary surface (whether the boundary reflects the phonons specularly or scatters them diffusely).

In a metal of high electrical conductivity, like copper, silver or gold of high purity, the *free electron contribution* to thermal conductivity is dominant. One can define a ratio, called the *Wiedemann-Franz* ratio, by the relation

$$W = K/\sigma T \tag{5.5.7}$$

In the free electron approximation this ratio should be a universal constant independent of temperature. For high purity metals at high temperature the experimental value for the Wiedeman-Franz ratio is close to the theoretical value.

The connection between thermal and electrical conductivity of a metal is not fortuitous. Both electrical and thermal conductivities arise from the scattering of electrons by phonons due to the electron-phonon interaction. In high purity metals this is the most dominant mechanism for scattering of electrons. However the scattering mechanism for electrical resistance involves the scattering of electrons through a large angle by phonons. On the other hand for thermal conductivity small angle scattering processes are important. This difference causes a deviation from the Wiedemann-Franz law as the temperature is lowered.

The electronic contribution to thermal conductivity is given by an expression of the same form as equation (5.5.4) except that c_{vel}, v and l now refer to the electronic specific heat, electron velocity and electron mean free path. The electron mean free path again varies as $1/T$ at high temperatures since the number density of phonons responsible for scattering electrons is proportional to $1/T$. The electronic specific heat is proportional to T. So at high temperature, the thermal conductivity of a metal does not vary much with temperature.

The thermal conductivity goes through a maximum at a temperature T_m and decreases to zero as the temperature is lowered to absolute zero. Defects in a metal, like substitutional impurities, dislocations and stacking faults scatter the electrons. These scattering processes gain in importance relative to phonon scattering as the temperature is lowered.

As the number density of defects is increased the maximum value, K_m, of thermal conductivity decreases and the temperature T_m at which the maximum occurs is raised. This is shown in Figure 5.4 for copper.

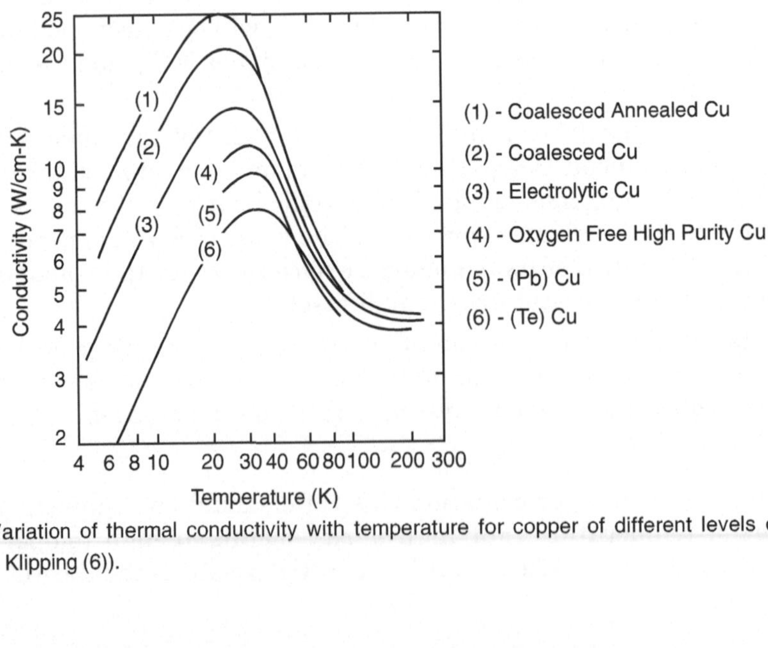

Figure 5.4: Variation of thermal conductivity with temperature for copper of different levels of purity (from I. Klipping (6)).

In an alloy like brass, substitutional atoms of zinc occupy the positions of copper atoms in the lattice. The fraction of sites occupied by zinc is large. In such a case the scattering due to substitutional atoms determines the mean free path even at high temperatures. The electron contribution to thermal conductivity becomes comparable to the phonon contribution. Both contributions taken into account lead to a much smaller value of thermal conductivity for brass than that for pure copper. The thermal conductivity of brass has also a much weaker dependence on temperature than that of pure copper. Figure 5.5 shows the thermal conductivity as a function of temperature for several materials.

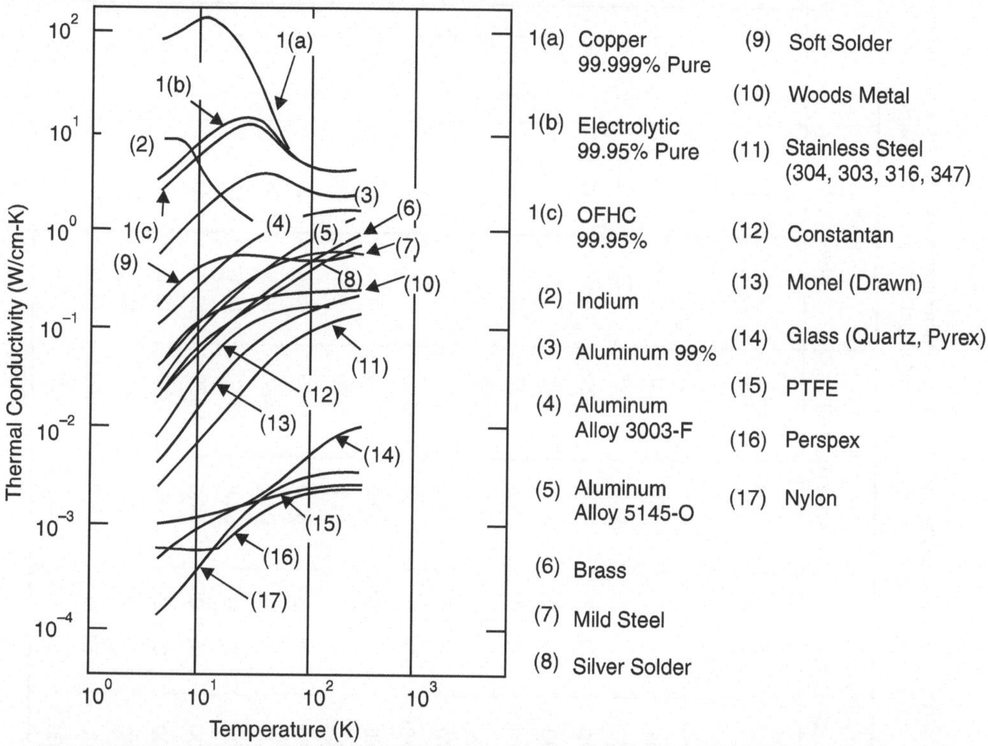

Figure 5.5: Temperature variation of thermal conductivity of different materials (from I. Klipping (6)).

One should take into account the dependence of thermal conductivity on temperature in calculating the heat flow through a material, the two ends of which are maintained at a temperature of 300 K and T. If the area of cross section of the material is A (in m^2), the length of the material is L (in m), the heat flux Q in Watts through the material from the warm to the cold end is given by

$$Q\,(L/A) = \int_T^{300} K(T)\,dT \tag{5.5.8}$$

Table 5.2: Thermal conductivity and thermal conductivity integrals for several materials (The thermal conductivity is in W/mK and the conductivity integral is in W/m)

Temp. (K)	*Copper (beryllium tough pitch)*		*Aluminium (6063 T5)*		*Low-carbon steel (C1020)*		*Stainless steel (304)*		*Monel (drawn)*		*Teflon*	
	K	$\int_{4.2}^{T} KdT$	K	$\int_{4.2}^{T} KdT$	K	$\int_{4.2}^{T} KdT$	K	$\int_{4.2}^{T} KdT$	K	$\int_{4.2}^{T} KdT$	K	$\int_{4.2}^{T} KdT$
4.2	1.9	0	34	0	3.0	0	0.24	0	0.43	0	0.046	0
10	4.8	19	86	360	11.5	43	0.77	2.9	1.74	6.3	0.096	0.44
20	10.6	95	170	1650	24.0	222	1.95	16.3	4.30	36.4	0.141	1.64
30	16.2	229	230	3650	32.0	502	3.30	42.4	6.90	92.9	0.174	3.23
40	21.0	415	270	6200	38.6	867	4.70	82.4	9.00	173	0.193	5.08
50	26.1	650	280	8950	47.6	1310	5.80	135	10.95	273	0.208	7.16
60	30.0	930	270	11700	53.6	1810	6.80	198	12.09	368	0.219	9.36
70	33.7	1250	248	14300	57.5	2360	7.60	270	13.06	513	0.228	11.6
80	37.0	1600	230	16700	60.0	2950	8.26	349	13.90	647	0.235	13.9
90	40.1	1990	222	19000	61.8	3350	8.86	436	14.63	791	0.241	16.3
100	43.0	2400	216	21100	62.9	4170	9.40	528	15.27	940	0.245	18.7
120	48.4	3300	207	25300	64.1	5450	10.36	726	16.26	1260	0.251	23.7
140	53.3	4320	201	29300	64.6	6750	11.17	939	17.34	1590	0.255	28.7
160	57.6	5440	200	33300	64.8	8050	11.86	1170	18.25	1950	0.257	33.8
180	61.5	6640	200	37300	64.9	9350	12.47	1410	19.02	2320	0.258	39.0
200	65.0	7910	200	41300	65.0	10200	13.00	1660	19.69	2710	0.259	44.2
250	72.4	11300	200	51300	65.0	13900	14.07	2340	21.02	3730	0.260	57.2
300	78.5	15000	200	61300	65.0	17200	14.90	3060	22.0	4800	0.260	70.2

In Table 5.2 the conductivity integrals from 4.2 K to 300 K for some cryogenic materials are given. The data are taken from the data book of I. Klipping (6). These data will be useful in making heat transfer calculations.

Here is an example of calculation of heat transfer by conduction. Let us consider a cryostat with a neck tube made of stainless steel SS304 with a diameter, d, of 10 cm, wall thickness, t, of 0.2 mm and length, L, of 20 cm. The warm end of the tube is at 300 K and the cold end is at 4.2 K. The area of cross section A for heat transfer is $A = \pi dt = 0.628 \text{ cm}^2 = 6.28 \times 10^{-5} \text{ m}^2$. Taking the conductivity integral from Table 5.2 the heat leak Q using equation (5.5.8) comes out to be 960.84 mW.

It is obvious that the smaller the wall thickness of the material and the smaller the conductivity integral, the lower is the heat leak. The tube will also have to carry a mechanical load. The wall thickness will also be determined by the mechanical strength required to carry the load. The material must also have sufficient impact strength to withstand sudden shocks. The mechanical properties of materials have been dealt within chapter 4.

6. EMISSIVE PROPERTIES

The amount of energy radiated by unit area of surface in unit time is given by

$$Q = \varepsilon\sigma T^4 \tag{5.6.1}$$

Here ε is the *emissivity* of the surface. It is a dimensionless number less than unity. σ is the *Stefan's constant* which has a value 5.67×10^{-8} W/m$^2K^4$ and T is the absolute temperature of the surface. This heat is propagated as electromagnetic waves and the energy will be absorbed by a second surface on which the radiation is incident. The absorption coefficient of a surface is equal to its emissivity. So the larger the value of ε the larger the quantity of heat absorbed.

If there are two parallel surfaces, one, which is warm, at a temperature T_W and the other, which is cold, at a temperature T_C, the heat transferred per unit area per unit time from the warm to the cold surface through radiation is

$$Q_r = \sigma\varepsilon_{\text{eff}}(T_W^4 - T_C^4) \tag{5.6.2}$$

The effective emissivity ε_{eff} is determined by the emissivities ε_C and ε_W of the cold and warm surfaces through the relation

$$1/\varepsilon_{\text{eff}} = 1/\varepsilon_C + 1/\varepsilon_W - 1 \tag{5.6.3}$$

The heat transfer by radiation takes place even in vacuum.

The emissivity of a surface depends on the condition of the surface. If a surface is highly polished, its emissivity is low (≈ 0.01) while for rough surfaces the emissivity is high. If the surface has a condensed layer of a gas, the emissivity depends on the thickness of the condensed layer. For a highly polished surface, the emissivity increases as the thickness of the condensate increases and reaches the emissivity value of the condensate for very large thicknesses of the condensate.

Table 5.3 lists the values of the emissivity of a few materials.

If the cold and warm surfaces are present in high vacuum (pressure 10^{-5} mbar or less) heat transfer by radiation plays a dominant role. Taking the emissivity as 0.1 and T_W = 300 K and T_C = 4 K, the heat flux per unit area by radiation has a value of approximately 48 W/m^2. To reduce the heat load one must use highly polished material with emissivity of the order of 0.01.

Table 5.3: Emissivities for a few materials (from Randall F. Barron (2))

Material	*Black body Source Temperature (K)*	ε_0 *at Absorptivity Temp. T*		
		Same as Source	*77 K*	*4 K*
Aluminium, annealed (Electropolished)	300	0.03	.018	.011
Aluminium with oxide layer				
0.25 μ thick	311	.06		
1.0 μ thick	311	.30		
7.0 μ thick	311	.75		
Aluminium with lacquer layer				
0.5 μ thick	311	.05		
2.0 μ thick	311	.30		
8.0 μ thick	311	.57		
Aluminium vaporised on 0.5 mil Mylar (both sides)	300		.04	
Brass				
Polished	373	.03		
Rolled plate	300	.06		
Shim Stock, 65 Cu/35 Zn	295	.35	.029	.018
Oxidized	373	.60		
Cadmium Electroplate (mossy)	295		.03	
Chromium plated on Copper	300	.08	.08	
Copper				
Black Oxidized	300	.78		
Scraped	300	.07		
Commercial Polish	300	.03		
Electrolytic, careful polish	295		.015	
Chromic acid dip	295		.017	
Polished	295		.019	

(Contd.)

Material	Black body Source Temperature (K)	ε_0 at Absorptivity Temp. T		
		Same as Source	77 K	4 K
Liquid honed	295		.088	
Electrolytic Polish	295			.0062
Carefully prepared surface	295	.018	.0082	.0050
Gold				
0.0015 in. foil (on Glass or Lucite)	295	0.02	.01	
0.0005 in. foil (on Glass or Lucite)	295		.016	
0.010 in Leaf (on Glass or Lucite)	295		.063	
Vaporised on both sided of 0.5 mil Mylar	295		.02	
Plate 0.0002 in. on SS (1% Ag in Au)	295		.025	
Plate 0.00005 in on SS (1% Ag in Au)	295		.027	
Iron				
Electrolytic	450-500	0.05-.65		
Cast. Polished	311	0.21		
Cast Oxidised	311	0.63		
Galvanised	365	0.07		
Stainless Steel Polished	373	0.08		
Stainless Steel, type 302	300		0.048	
Lead				
0.004 in. Foil	295	0.05	0.036	0.011
Grav. Oxidised	295	0.28		
Magnesium	295	0.07		
Mercury	273-373	0.09-0.12		
Nickel				
Electrolytic	295	0.04		
0.004 in. Foil	295		0.022	
Electroplated on Copper	300		0.03	

6.1. Multilayer Insulation

We may interpose n radiation shields between the warm and the cold surfaces as shown in Figure 5.6. Then the heat flux per unit area from the warm to the cold surface becomes reduced to

$$Q_r = [1/(n+1)]\, \sigma\varepsilon_{\text{eff}}(T_W^4 - T_C^4) \tag{5.6.4}$$

For simplicity it is assumed that the effective emissivity between a pair of adjacent surfaces is a constant. It is seen that the interpolation of shields reduces the heat transfer by radiation.

If one uses thin aluminised mylar (thickness about 6 μm) then one can wind the thin film in several layers around the cold wall. However as the density of layers is increased solid contact conduction will come into play. One can use spacers between the layers to reduce the area of contact between the layers and thus reduce solid contact conduction.

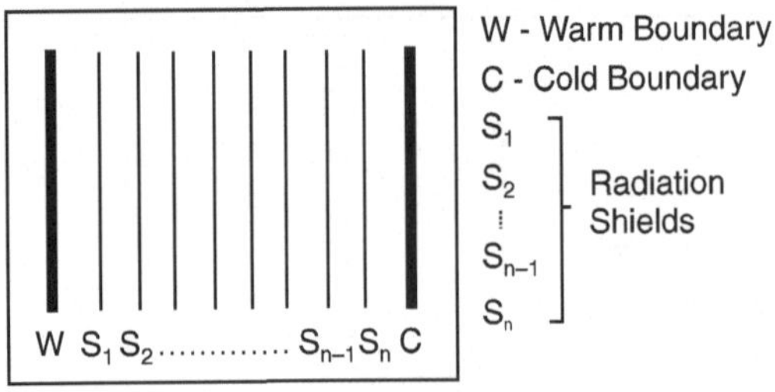

Figure 5.6: Interposition of *n* radiation shields between the end warm and cold surfaces.

Even so the optimum density of layers is in between 10 to 20 per cm. With such tightly wound layers it will be difficult to evacuate the space between the warm and cold walls. Air trapped between the layers will find it difficult to get out. To avoid this, holes are punched in the aluminised mylar sheet. In winding the layers care must be taken to see that the holes do not line up to provide an unobstructed passage for radiation to travel from the warm to the cold wall.

It is necessary to warm the insulating layers by blowing hot air on the inside of the container during the evacuation process to facilitate the removal of gases adsorbed on the surface of the aluminised mylar layers. It is the usual practice to keep some activated charcoal in contact with the bottom outer surface of the cold wall to adsorb residual gases. When liquid helium is collected in the vessel the charcoal will cool and cryosorb the gases to maintain a high vacuum.

The layers of super-insulation are bunched together and anchored to vapour shields at different points in the neck of the vessel. Cold vapour evaporating from the vessel will be used to cool the layers of super-insulation to temperatures intermediate between the warm and cold walls.

REFERENCES

1. B.A. Hands, "Cryogenic Engineering", Academic Press, New York, 1986.
2. Randall F. Barron, "Cryogenic Systems", 2nd edition, Oxford University Press, New York, 1985.
3. Thomas M. Flynn, "Cryogenic Engineering", Marcel Dekker, Inc., New York, 1997.
4. E.S.R. Gopal, "Specific heats at low temperatures", Plenum Press, New York, 1966.
5. K.D. Timmerhaus and T.M. Flynn, "Cryogenic Process Engineering", Plenum Press, New York (1989).
6. I. Klipping, "Datensammlung zur Kryotechnik", VDI-Bildungswerk Verein Duetscher Ingenieure, BW2407 (1998).

CHAPTER 6

Principles of Cryostat Design

1. INTRODUCTION

Now-a-days one can buy commercial systems for carrying out any type of measurement at low temperatures. These systems allow one to make simultaneous measurements on a number of samples of a variety of properties with a single measuring system. Such systems are costly but convenient. There are many such property measurement systems installed in a variety of institutions in the country. For university users facilities are available in the UGC-DAE Consortium for Scientific Research for measuring resistivity, specific heat and magnetisation in magnetic fields which can be varied upto 14 Tesla.

However it is good to have a knowledge of the principles used in designing cryostats. In this country many university groups may not be able to afford commercially available cryostats. They may like to make a cryostat of their own for limited use. It is hoped that this Chapter will help such groups.

An appropriate cryostat will have to be designed for carrying out low temperature measurements of different properties. Before embarking on the design the following questions should be answered:

(1) In what temperature range will the cryostat operate; (*a*) above 77 K, or (*b*) above 4.2 K or (*c*) above 1.8 K?

(2) Should the cryostat be of the bath type, continuous flow type or should it be cooled by a closed-cycle refrigerator?

(3) What is the nature of measurement to be done and the size of the sample to be used?

(4) Would one like to perform a fixed temperature or a variable temperature measurement?

Once these questions are answered, the design of a suitable cryostat can be made with the available materials in the market and the technical expertise in the laboratory.

2. PRECAUTIONS AGAINST DIFFERENTIAL CONTRACTION

The most commonly used materials in the construction of cryostats are stainless steel, copper and brass. All these materials do not become brittle at low temperatures. Stainless steel has a large strength to weight ratio and one can use thin walled tubes for construction. This reduces the weight of the cryostat. However for the sample holder and the sample chamber one uses copper. The large thermal conductivity of this material makes for a greater uniformity of temperature. Brass is also used occasionally for making vacuum cans for the sample holder. To reduce the radiation heat load, one should plate the low temperature parts with gold. This is especially recommended for very low temperature work. For electrical insulation one may use teflon in parts of the cryostat.

We see that in the construction of a cryostat one may have to use dissimilar materials. In such a case one factor that requires careful consideration is the relative thermal expansion of the different materials. The cryostat will be subjected to warming and cooling repeatedly. Differential contraction of dissimilar materials will then lead to opening of joints leading to leaks.

The cumulative thermal expansion for unit length is given by

$$\Delta L/L_0 = \int_0^T \alpha \, dT \tag{6.2.1}$$

This data for various materials is given in Table 6.1. L_0 is the length of the sample at 0 K. The contraction of stainless steel from 300 to 10 K is less than that of copper, which in turn is less than that of brass.

Table 6.1: Cumulative thermal expansion for unit length, $\Delta L/L_0$, for different materials as a function of temperature *T*. The values have to be multiplied by 10^{-5} (from H. Walter (1))

Temp.	*Material*					
[K]	*Stainless Steel*	*Brass*	*Copper*	*Aluminium*	*Nylon*	*Teflon*
0	0	0	0	0	0	0
20	0	1	1	0	10	30
40	0	4	4	3	37	80
60	4	16	12	9	81	140
80	13	34	26	24	142	210
100	32	58	45	46	217	290
120	49	85	67	71	301	380
140	72	115	92	104	393	480
160	97	147	120	138	493	600
180	122	180	149	175	600	740
200	150	215	179	214	719	900
220	179	249	210	254	845	1110
240	208	283	241	295	981	1310
260	240	320	273	339	1109	1450
280	272	357	305	384	1284	1535
300	304	397	337	418	1450	1600

In making a joint between such dissimilar metals one should take care to see that the material with the higher contraction is outside the material with the lower contraction as shown in Figure 6.1(*a*). When the joint is cooled the outer tube will contract more and squeeze the inner tube.

If the tube, made of material with a smaller contraction, is put outside a tube made of material with a larger contraction, there is a danger of solder giving way and the joint leaking at low temperatures. Similarly in using a flange of stainless steel it is better to use a brass bolt to tighten the flange with a stainless steel washer as shown in Figure 6.1(*b*). Brass contracts more than stainless steel on cooling and this makes the contact tighter on cool down.

Generally epoxies like Araldite contract much more than a metal. In making electrical feed throughs one uses an epoxy to isolate electrically the metal leads from the flange. As the feed-through is cooled the epoxy contracts more than the metal. If the epoxy is filled in a hole in the flange through which the electrical lead is brought out, it will crack from the wall of the flange when the lead is cooled. The correct way to make a feed through is to braze or solder a thin walled SS tube to the flange. The lead is brought out through the tube and the epoxy is applied to cover the outside of the thin walled tube. The epoxy will then squeeze the tube when cooled and this will preserve the integrity of the seal. Often inadequate attention to such detail will lead to cold leaks and loss of valuable time in detecting such leaks.

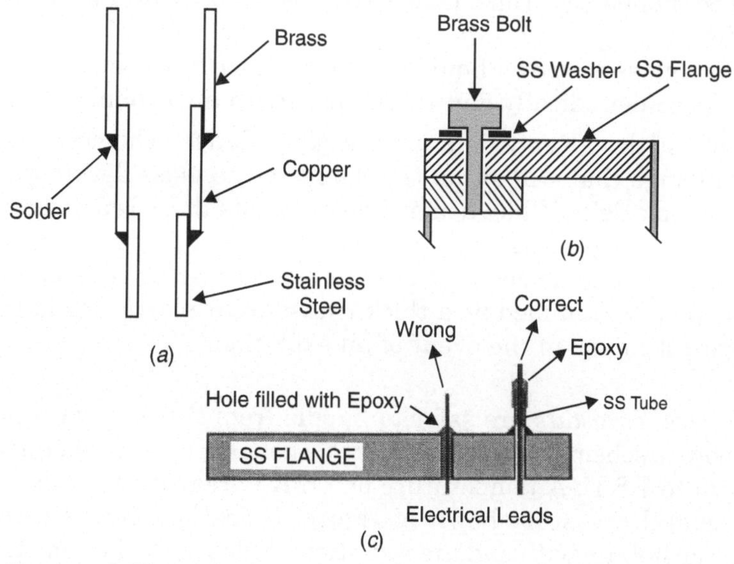

Figure 6.1: Taking care of differential contraction effects in joints of dissimilar materials (from H. Walter (1)).

Sometimes the sample chamber is connected by several tubes to the top flange. In low temperature cryostats the length of such tubes is usually large. If the tubes of dissimilar materials are used, one will contract more than the other and this will lead to large stresses at the joints of the tubes. One must pay particular attention to avoid such stresses.

Similarly in taking thin wire leads from the sample chamber to the top flange sufficient slack is to be allowed in the wires. Otherwise differential contraction will make the wires snap when the cryostat is cooled.

3. BATH CRYOSTATS

Cryostats using liquid cryogens can be conveniently classified in one of two types: bath cryostats or continuous flow cryostats. In the former the entire experimental set up is immersed in a bath and is suspended from a top flange. In the latter, the liquid cryogen is sucked by a vacuum pump and flows through the parts to be cooled.

One can make a bath cryostat either out of stainless steel or out of glass. For a beginner in low temperature experimentation, a *glass bath cryostat* has the advantage that one can see through the walls of the cryostat. A bath type cryostat made of glass is shown in Figure 6.2. For liquid nitrogen the glass cryostat can be made of pyrex glass. The liquid cryogen is contained in a double walled glass dewar. The space between the walls can be evacuated to a high vacuum and sealed. To reduce evaporation loss from the dewar, one may also coat the inner wall of the outer tube and the outer wall of the inner tube with silver leaving two vertical slits opposite each other for viewing.

Helium gas slowly permeates through the walls of the glass dewar and the vacuum will be spoiled with time. Lead glass has a much lower diffusion coefficient for helium gas and so glass dewars for helium are also made of lead glass. It will be better to make a glass dewar for liquid helium work with provision for re-evacuating the vacuum space periodically and resealing it with a mechanical seal.

The top of the dewar ends in a ground glass flange. The cryostat parts hang from a metal flange that will sit on the flange of the glass dewar with an O-ring seal to make it leak tight. While using glass cryostats there is a danger of the glass dewar exploding or imploding. If the glass dewar is not properly annealed there is a danger of this happening sooner than later. As a matter of safety, a glass bath cryostat should be surrounded by a thick transparent plastic shield to protect the personnel against injury in the event of an explosion.

Stainless steel cryostats are safer to handle from the long term point of view. Figure 6.3 shows a schematic diagram of such a general-purpose metal bath cryostat for work down to 1.8 K. A temperature of 1.8 K is reached by pumping on liquid ^{4}He. In the metal cryostat shown in figure, there is a liquid nitrogen shield surrounding the helium bath and the radiation baffles in the helium dewar prevent radiation from the top of the dewar from reaching the helium bath.

The sample chamber in the dewar can be evacuated or filled with exchange gas of helium at about one mbar pressure. If one wants to measure the temperature variation of a property like electrical resistivity, then one can evacuate the sample chamber after the sample has reached the lowest temperature and then heat the

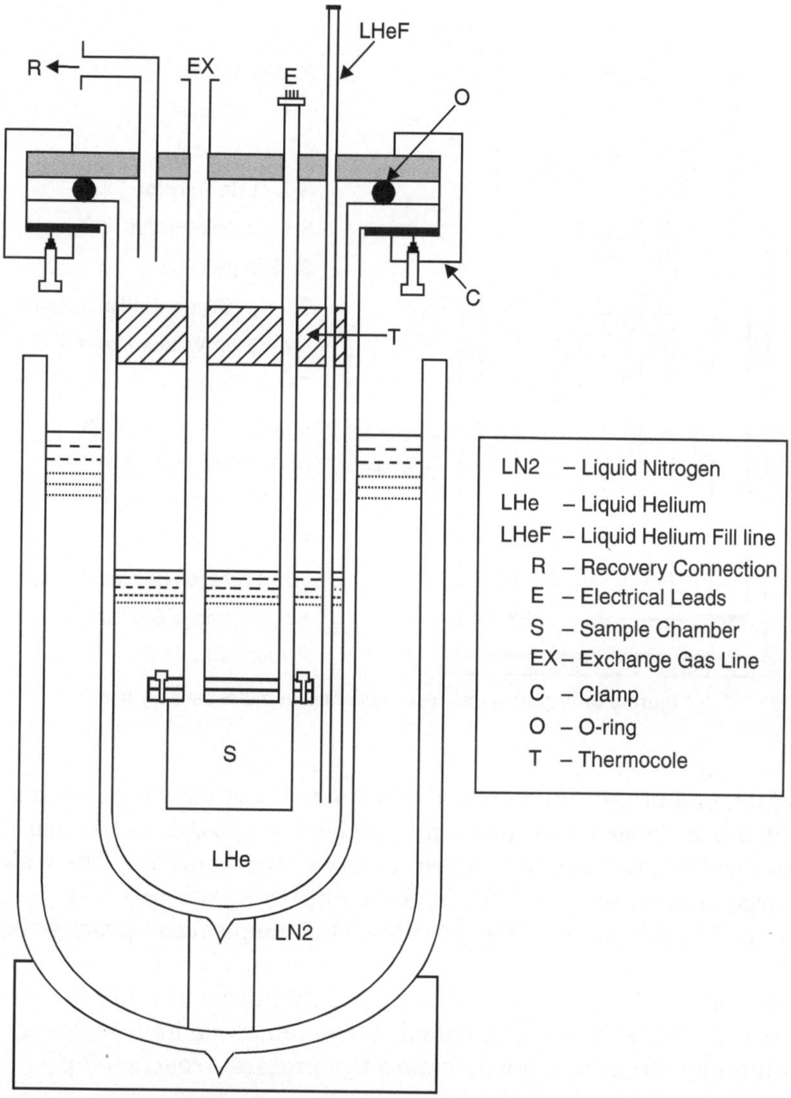

Figure 6.2: A schematic of a glass dewar

sample at a slow rate. It should be noted that the electrical leads should be made of thin (SWG 40) copper or manganin wires and the wires should be anchored at a point at 4.2 K as shown. This will reduce the heat leak through the leads.

Such a bath cryostat, about a meter in height with a capacity of ten litres of liquid helium, can be fabricated at a cost of about Rupees fifty thousand. It is important to have a bursting disc and a pressure relief valve on the outer vacuum chamber. In case there is a leak of cryogen from the nitrogen or helium dewars, the liquid will evaporate rapidly, building up pressure in the outer vacuum vessel. The pressure relief valve and the bursting disc prevent the pressure from reaching such a high value as to cause the outer vessel to explode.

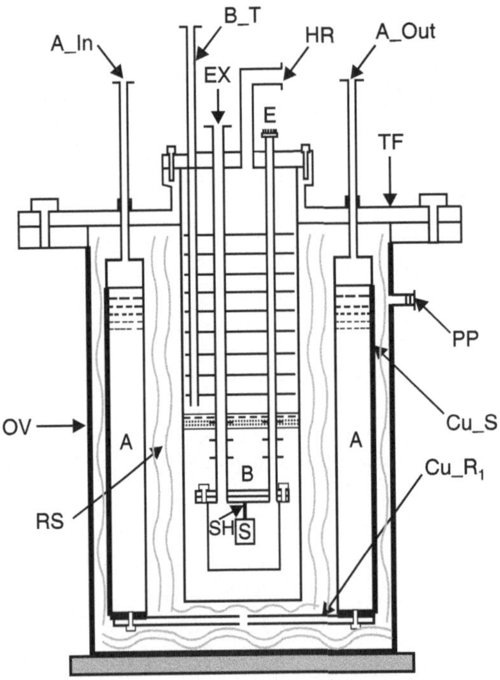

A - LN2 Vessel
A_In - LN2 Inlet
B - LHe Vessel
B_T - LHe Transfer Tube
SH - Sample Holder
S - Sample
Cu_R_1 - Copper Radiation Shield
OV - Outer Vacuum Vessel
TF - Top Flange
E - Electrical Feedthrough
HR - Helium Recovery
Cu_S - Copper Conducting Shield
PP - Pumpout Port
RS - Radiation Shields (Super insulation)
EX - Exchange Gas Connection
A_Out - LN2 Vent

Figure 6.3: A general-purpose liquid helium metal bath cryostat

In the construction of helium and nitrogen vessels one uses a stainless steel sheet (SS 304) of suitable thickness. From the point of view of reducing heat leak, one should use a sheet of as small a thickness as possible. In case the helium evaporates fast the pressure in the helium dewar will build up. The wall thickness must be large enough to withstand a sufficient over-pressure. At the same time the heat leak should be as small as possible. One makes a compromise between the two conflicting requirements and chooses a wall thickness to withstand an overpressure of about 5 psi. To provide for safety in case the pressure crosses this value, a pressure relief valve is provided on the top of the helium dewar which will release helium gas from the dewar before the pressure reaches 5 psi.

When the pressure inside the dewar is more than the pressure in the outer chamber the wall of the dewar experiences a tensile force. On the other hand in the reverse situation, the wall of the dewar will experience a compressive force. For the same thickness, the wall of the dewar can withstand a higher tensile force than a compressive force. This implies that the above dewar may not withstand even a differential pressure of a few psi under compression. If the helium dewar is evacuated while the outer vessel is at atmospheric pressure the inner dewar will crumple. This is known as buckling. Before evacuating the inner helium dewar one should make sure that the outer vessel is evacuated. This should be taken care of during the leak checking on the dewar.

The wall thickness for the cylinders and the dished ends of the dewar both for bursting as well as buckling can be calculated using the ASME pressure vessel codes, Section VIII (Unfired Pressure Vessels). The details are presented in *Appendix*.

3.1. Cool Down of a Bath Cryostat

The helium dewar must be first cooled to near 77 K before starting to fill liquid helium in it. The liquid nitrogen dewar is filled first and the helium dewar is allowed to cool by radiation. This will take several hours but will not contaminate the helium dewar.

For a more rapid cool down one may fill the helium dewar with some liquid nitrogen. Once the helium dewar is cooled to 80 K, liquid nitrogen in the dewar should be *completely* boiled off before one starts cool-down of the inner dewar with liquid helium. This can be done by putting a heater inside the helium dewar and switching it on. If liquid nitrogen is left in the dewar before helium filling is started, a lot of liquid helium will be used for cool down to 4.2 K. This is because the thermal capacity of liquid and solid nitrogen is high compared to the thermal capacity of the material of the dewar.

Another point needs to be made about the cool down with helium. One may use only the latent heat of vaporisation of helium for the cool down or one may use the complete enthalpy of helium from 4.2 K to 300 K. For liquid helium the latent heat of vaporisation at 4.2 K is smaller by a factor of 40 than the total enthalpy from 4.2 to 300 K. If one uses a rapid cool down of the cryostat with liquid helium a lot of cold helium vapour will exit the dewar and the consumption of cryogen will be excessive. If the total enthalpy of cooling is used, the exiting helium gas will be near 300 K and the consumption of cryogen will be minimal. Though the cooling will take a longer time to achieve, it is preferable to use a slow cool down using the total enthalpy of helium. Table 6.2 gives the amount of cryogen required to cool down 1 kg of different materials.

Table 6.2: Amount of Cryogen in litres required to cool 1 kg of the metal in different temperature ranges. (*a*) Cooling only using the Latent heat of Vapourisation (*b*) Cooling using the total Enthalpy of the cryogen from T to 300 K.

Cryogen	*From* (K)	*To* (K)	*Aluminium* a	*Aluminium* b	*Copper* a	*Copper* b	*Stainless Steel* a	*Stainless Steel* b
NITROGEN	300	77	1.0	0.63	0.46	0.28	0.53	0.33
HELIUM	77	4.2	3.2	0.20	2.20	0.16	1.40	0.10
HELIUM	300	4.2	66	1.60	32.0	0.80	34.0	0.80

A bath cryostat has the disadvantage that it can only be used in the vertical orientation. The consumption of cryogen is more the higher the temperature at which an experiment has to be done.

4. CONTINUOUS FLOW CRYOSTATS

When a sample in a small volume has to be cooled and when the experiments have to be done at variable temperature, the continuous flow cryostat is eminently suitable. Figure 6.4 shows a general-purpose continuous flow cryostat.

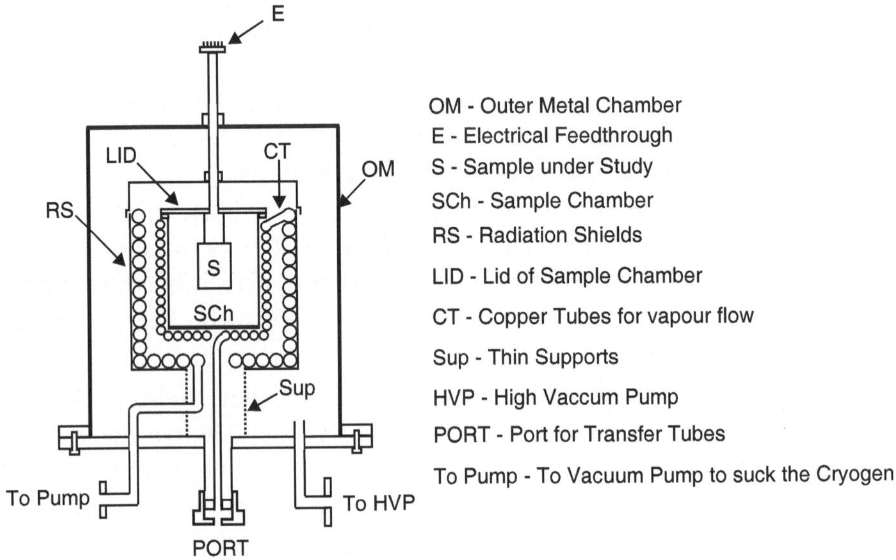

Figure 6.4: A schematic diagram of a general-purpose continuous flow cryostat

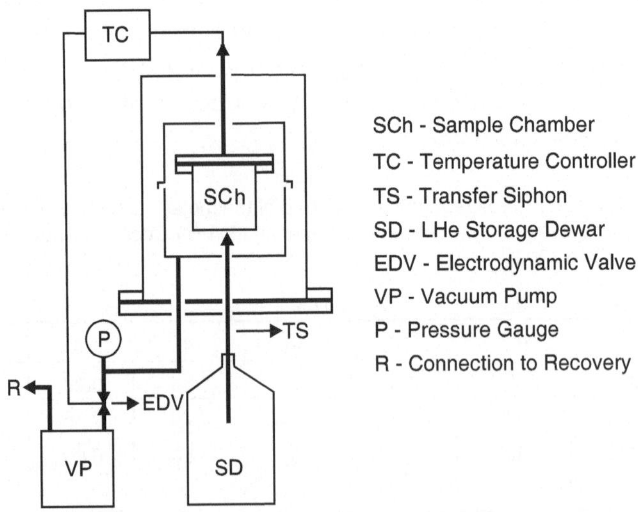

Figure 6.5: Flow diagram in a continuous flow cryostat

Figure 6.5 shows how the continuous flow cryostat can be used. A vacuum pump sucks the cryogen from the storage dewar through the vertical transfer tube.

The cold vapour from the storage dewar passes through the coil wound on the sample chamber and cools the chamber. A radiation shield surrounds the sample chamber and is cooled by the cold vapour flowing through a coil attached to it. The coil wound round the radiation shield is made of larger diameter tubing than the coil wound on the sample chamber to take care of the increase in specific volume of the vapour on heating. The vapour leaves the shield and reaches the vacuum pump. The rate of flow of the cryogen can be adjusted by a needle valve at the inlet to the pump to ensure that the full enthalpy of the vapour is used in cooling the sample chamber. By finely controlling the flow with an electrodynamic valve the temperature of the specimen chamber can be controlled to an accuracy of 10 mK or better.

The sample is mounted in the sample chamber, which can be filled with exchange gas at low pressure to maintain uniformity of temperature. To change the sample, the flow of the cryogen is shut off, and, after the sample chamber has warmed up to room temperature, the old sample is removed and the fresh sample loaded from the top into the sample chamber. The consumption of cryogen as a function of temperature of operation of the sample chamber is shown in Figure 6.6. The higher the temperature of the sample chamber, the smaller the consumption of cryogen. The continuous flow cryostat enables one to perform variable temperature experiments with economic consumption of cryogen.

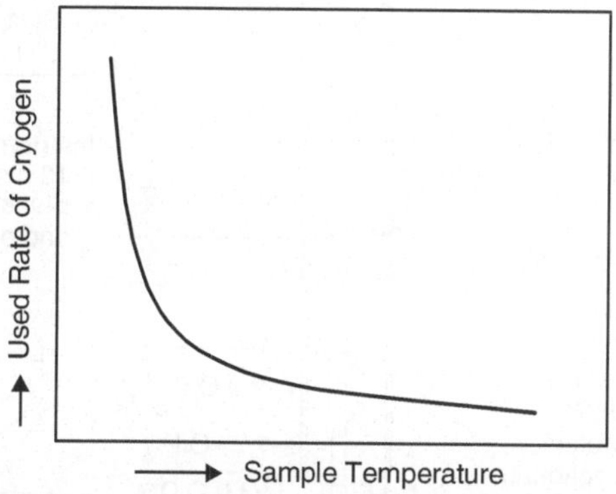

Figure 6.6: Schematic showing the consumption of cryogen as a function of temperature in a continuous flow cryostat

The continuous flow cryostat can be mounted in any orientation. The continuous flow cryostat consumes less cryogenic fluid, the higher its temperature of operation. The temperature can be easily controlled by controlling the flow. However its temperature stability is good only when the flow through the cryostat is a single-phase flow. When the flow is two phase flow large temperature fluctuations occur. For example, with liquid helium, it is difficult to achieve stable temperatures in the range 4.2 to 15 K.

Since a continuous flow cryostat can be operated in any orientation, such cryostats have been built for a variety of applications such as cooling a sample for electron microscopy, X-ray diffraction, optical and magnetic studies etc.

5. CLOSED CYCLE REFRIGERATOR CRYOSTAT

Where handling of liquid nitrogen and liquid helium is not desired or obtaining these liquids is difficult, a mechanical (closed cycle) refrigerator can be used to cool down the samples. The most common commercial refrigerators make use of a two-stage GM cycle using high purity helium gas as the working fluid. The cryocoolers are discussed in Chapter 3.

Figure 6.7 shows the cold head of a closed cycle refrigerator (with mechanically driven pistons), to which is attached a vacuum jacket and an adapter containing the electrical feed throughs and a pressure relief valve. The vacuum jacket surrounds the cold head. With a two-stage cryocooler, the second stage cold head acts as a sample mount, while the first stage is used for mounting a radiation shield surrounding the sample. The cold finger can be oriented in any direction in

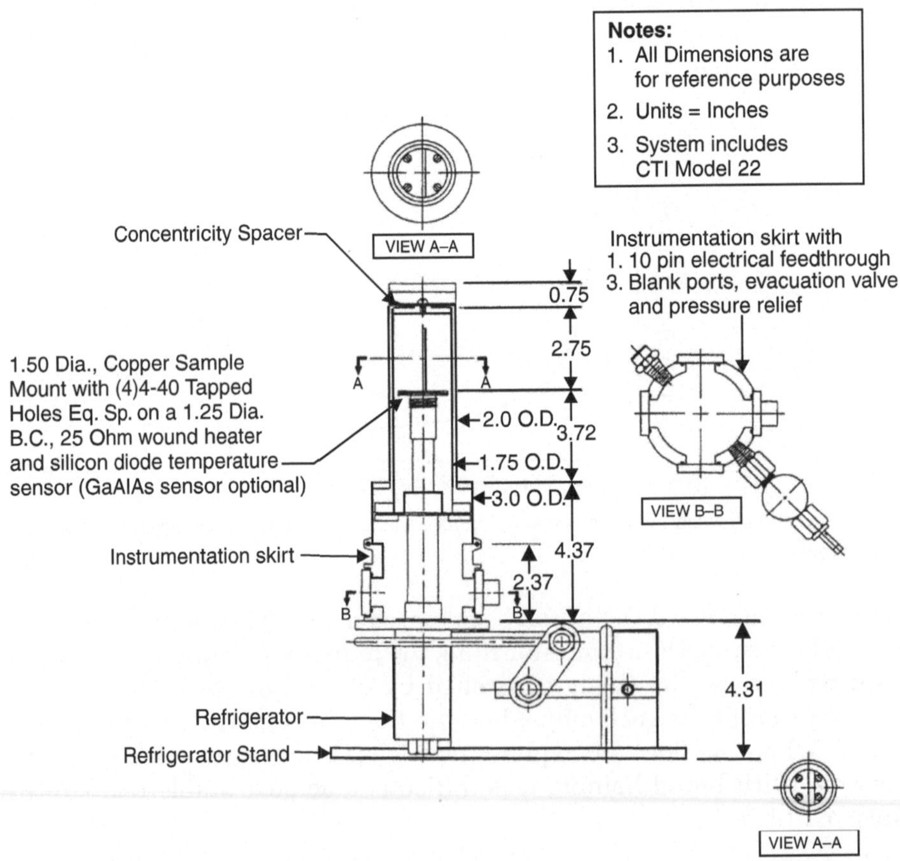

Figure 6.7: Closed Cycle Refrigerator Cryostat (*Courtesy:* M/s Janis Research Company Inc.)

the case of GM or Stirling type Cryo-refrigerators. However, for the Pulse Tube based systems, the lowest temperature would be reached when the sample mount is at the lowest position.

A typical system will reach the lowest temperature in about one hour. This will be affected by the sample size and the associated holder. Some units may exhibit a small temperature fluctuation at the lowest temperature, which can be eliminated by using an automatic temperature controller to operate at a slightly higher temperature.

With a multi-stage closed cycle refrigerator using suitable regenerator materials one can reach a temperature of 4.2 K or below. With such a closed circuit refrigerator one can make measurements on small quantities of samples down to liquid helium temperature. Using high T_c superconducting leads one can cool a small superconducting magnet to 4.2 K and make measurements on these samples in magnetic fields. However larger property measuring systems only use a bath cryostat.

6. SEALS AT LOW TEMPERATURE

All low temperature experiments are done in vacuum. The top flange of a cryostat is at room temperature and one can use a neoprene rubber gasket for sealing. On the other hand the sample chamber will be at low temperature and needs to be sealed against leaks using special material and design.

Many commonly used sealing materials at room temperature become brittle at low temperatures. Teflon gaskets can be used for sealing down to 77 K. With such gaskets and flanges made of materials like copper and aluminium, the mating flanges have to be thick to maintain the necessary pressure for sealing. Teflon gaskets are not therefore suitable for use at low temperatures.

The most useful sealing material at low temperature is the soft metal indium. This metal forms a good sealant because it deforms plastically to fill the O-ring groove completely. When making an indium seal it will be preferable to have the mating flanges made of the same material say stainless steel. Figure 6.8(*a*) to (*c*) shows various possible groove profiles for indium seals. The profiles are in increasing order of complexity. Figure 6.8(*d*) shows a simple indium O-ring seal for joining two tubes. This can be easily fabricated and assembled.

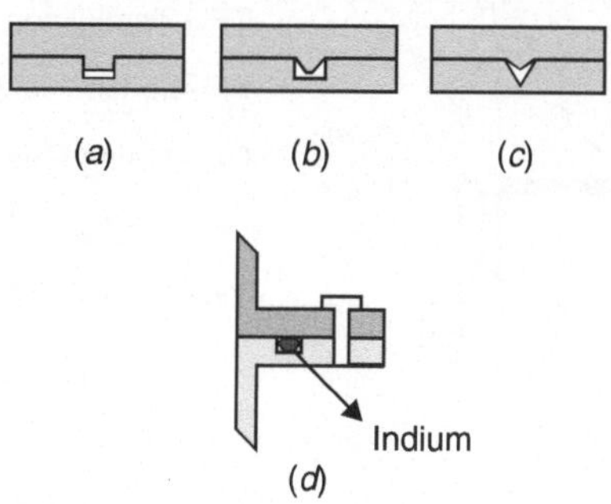

Figure 6.8: A simple Indium seal (from (1))

One may use the differential contraction of dissimilar metals to make a vacuum tight joint. Such a seal is shown in Figure 6.9. The sealing action arises here because aluminium shrinks more than copper on cooling. This type of sealing is useful if one has to have demountable windows on the cold inner dewar of a bath cryostat.

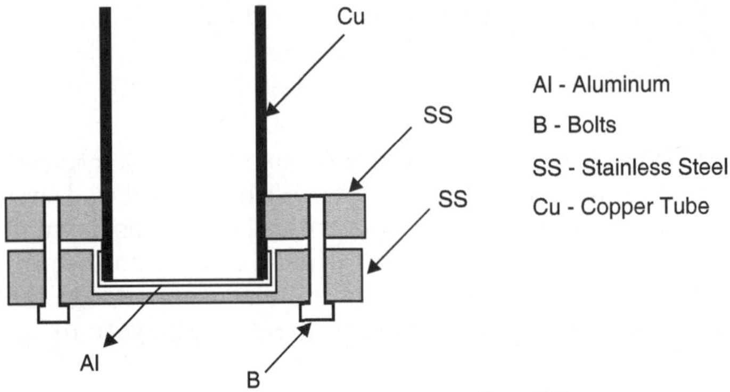

Figure 6.9: A contraction seal (from (1))

7. OPTICAL WINDOWS

For optical experiments light from an external source will have to pass through the cold sample in a dewar. Light will enter through two windows, one on the outer vacuum shield at room temperature and the other on the cold wall of the helium dewar. Light will also exit through similar windows. While the first window at room temperature can be fixed with a conventional viton O-ring seal, the second window requires special construction.

Figure 6.10(*a*) shows one method of fixing the optical window using an indium seal. The indium metal ring on the front of the window serves to distribute the

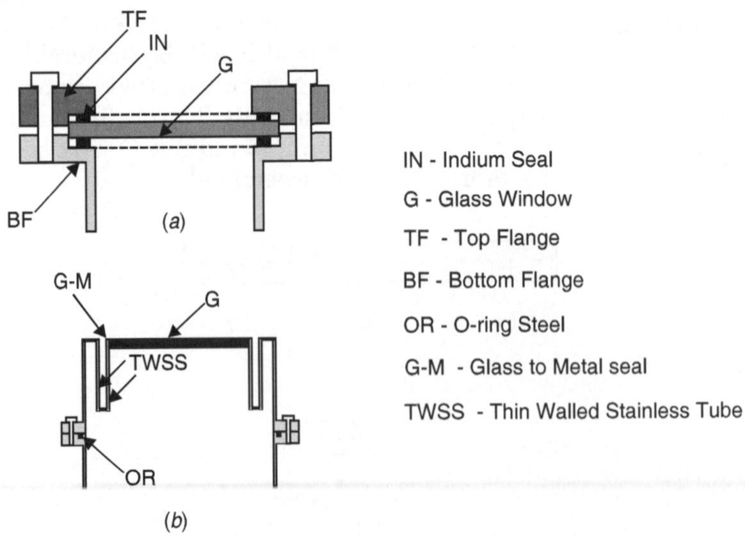

Figure 6.10: Optical windows for cryogenic work

pressure uniformly on the window material when the bolts are tightened on the flange. It does not prevent leaks. It is the indium O-ring seal behind the window material which prevents leaks. In this arrangement the window material is subjected to uniaxial pressure. This is unacceptable if polarised light is to be used for the studies.

One should then have a window with no load on it. Such an arrangement is shown in Figure 6.10(*b*). The window is fused to a tubular fixture using a glass to metal seal. This tubular fixture is joined to a concentric thin walled tube, which is then brazed on to the connecting tube on the flange. In this arrangement tightening the bolts on the flange does not cause any stress on the window material.

8. MOUNTS FOR THERMOMETERS

In any measurement one would like to measure the temperature of the sample. The different sensors that can be used to measure the temperature will be discussed in Chapter 7. The most commonly used thermometer is the resistance thermometer.

A resistance thermometer is fixed to the body of the sample holder by an epoxy, which is electrically insulating but thermally conducting. Any heat leaking into the thermometer from a higher temperature will warm up the thermometer. The reading of the thermometer will be erroneous.

To avoid such a heat leak the thermometer leads are put in good thermal contact with the body of the sample holder before they are led out. This ensures that the thermometer and sample holders are at the same temperature. Epoxy has a high emissivity and will absorb radiation from the high temperature parts of the cryostat to which it is exposed. To prevent such heat from reaching the thermometer, the thermometer is put in a groove in the body of the sample holder and covered with epoxy. A radiation shield is fixed on the epoxy and anchored to the sample holder. This is shown in Figure 6.11(*a*).

Another way of mounting a thermometer is to put it in a copper tube, which is evacuated and filled with helium exchange gas at 1 mbar pressure. This tube is soldered to a copper plate, which is tightly screwed to the body of the sample holder. This is shown in Figure 6.11(*b*).

In case one uses a vapour pressure thermometer a possible arrangement is shown in Figure 6.12. Here a small hole is bored in the wall of the sample chamber or sample holder, as the case may be, to hold a few mm^3 of the liquid used for vapour pressure thermometry. This hole is connected by a narrow stainless steel capillary to the room temperature flange. The tube can be evacuated, purged and filled with the pure gas to be used in the vapour pressure thermometer. One should use a fairly long length of capillary so that the heat leak into the sample holder is small.

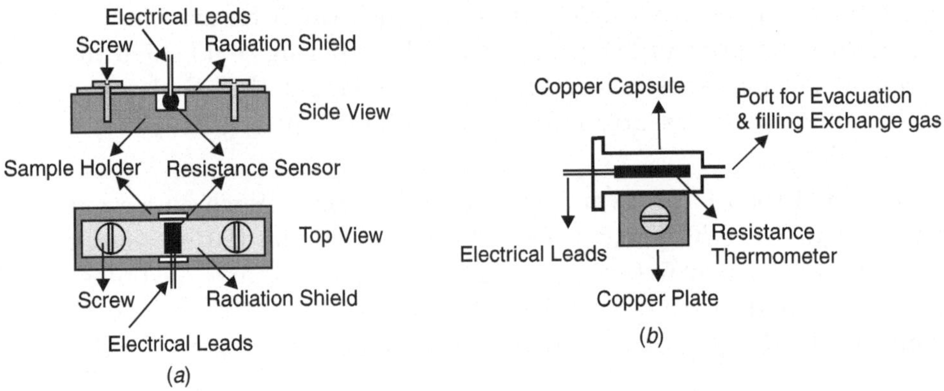

Figure 6.11: Thermometers mounting Arrangements (from (1))

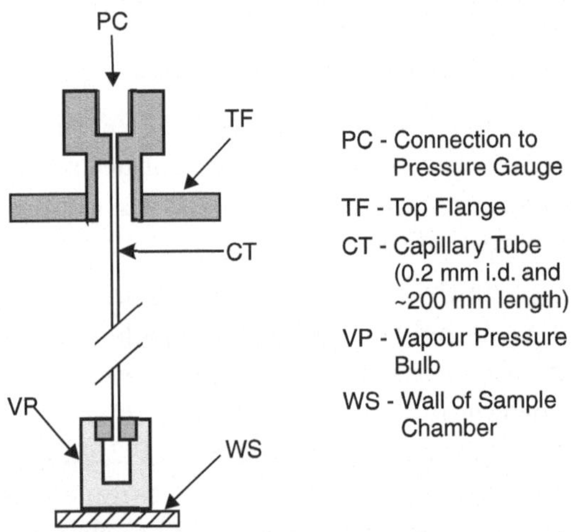

Figure 6.12: Vapour pressure bulb and capillary (Drawing not to scale)

9. DIFFERENT TYPES OF COMMERCIAL CRYOSTATS

Commercial experimental cryostats are available for different applications. M/s Janis Research Company Inc. is one of the major suppliers of such cryostats and the figures of some of their cryostats are shown below.

Figures 6.13(*a*), (*b*), and (*c*) show a detachable tailed dewar with liquid nitrogen reservoir for liquid helium experiments, an optical exchange gas cryostat, a variable temperature magnet cryostat respectively. For further details, the reader should refer to the technical literature available from the company.

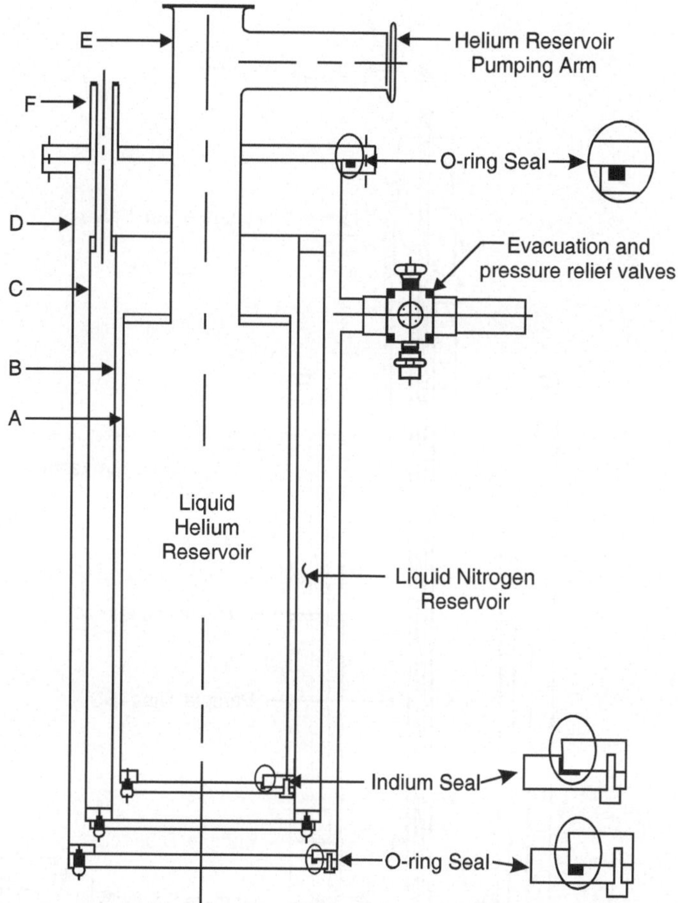

Figure 6.13 (a): A detachable tailed dewar with liquid nitrogen shield for liquid helium experiments (from Janis Research Co.)

10. MEASUREMENTS WITH CLOSED CYCLE REFRIGERATORS

When one uses a closed cycle refrigerator, the sample holder should be attached in good thermal contact with the cold head of the cryocooler. One can design different sample holders for different measurements. The radiation shield for the sample holder is usually anchored to the first stage of the cryocooler. Since the space available within the outer vacuum mantle of a cryocooler is limited, the size of the sample holder becomes a constraint. Since most of the cryocoolers can be mounted in any orientation, one can design suitable sample holders for different measurements, optical, magnetic, Mossbauer etc.

The vibrations in a closed cycle refrigerator will vitiate measurements, for example in a Mossbauer set up or in a vibrating sample magnetometer. Suitable precautions will have to be taken to reduce the vibration level.

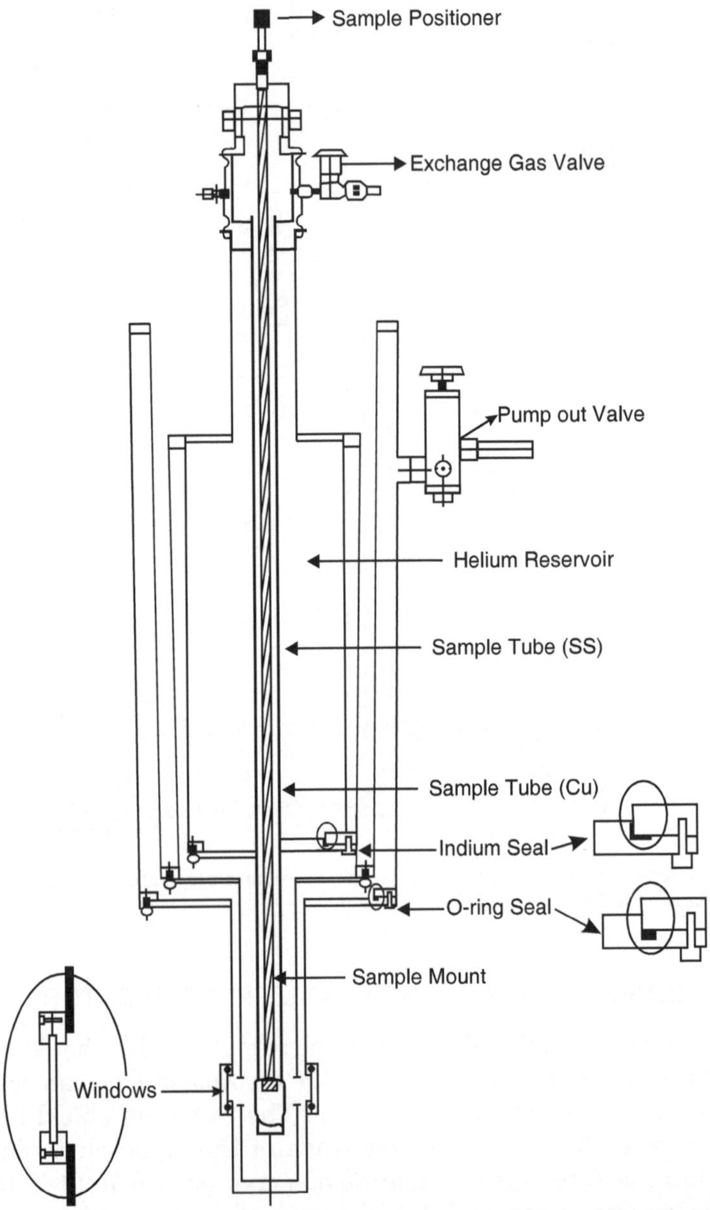

Figure 6.13(b): An optical Cryostat for liquid helium experiments (from Janis Research Co.)

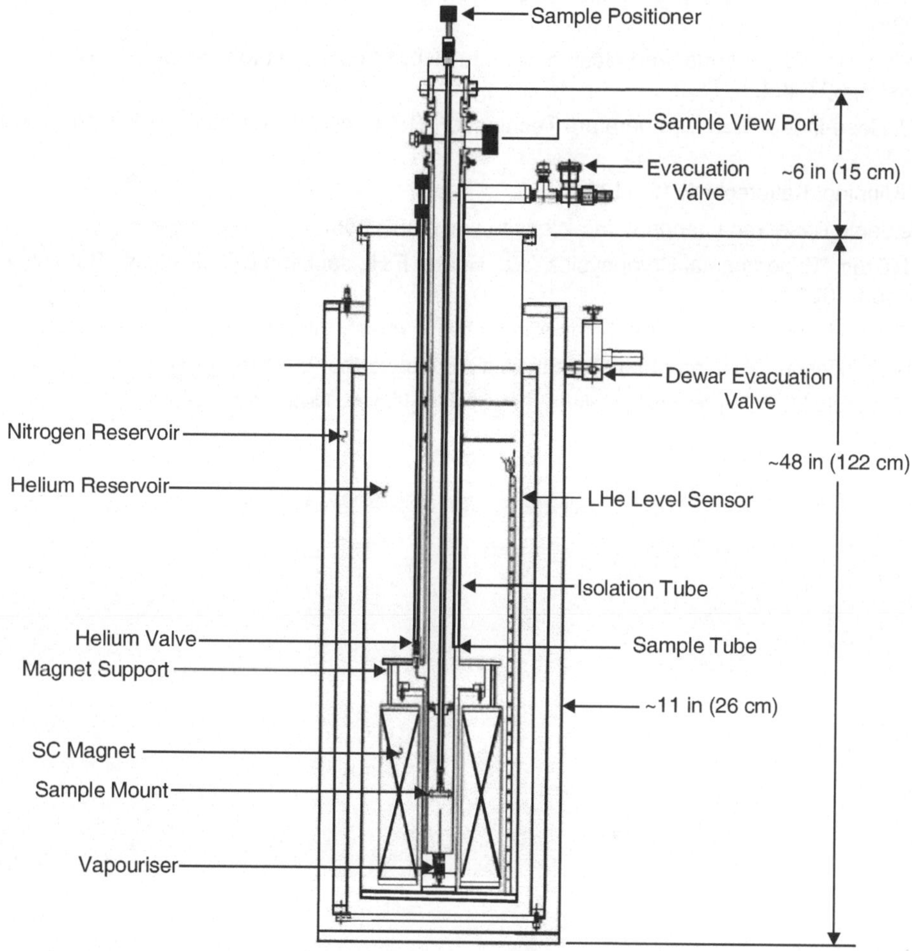

Figure 6.13(c): A Superconducting magnet Cryostat for variable temperature and liquid helium experiments (from Janis Research Co.)

11. CONCLUSION

In this chapter, some general principles of cryostat design have been mentioned. Many other tips will be found in the books given below. It is believed that these tips will be useful even for the users of commercially available cryostats in case the cryostats go bad and need to be serviced.

Many cryostats for different measurements are described in the literature. It is not possible to describe such cryostats in detail within the purview of the book. The reader should consult the literature and decide on a suitable design of a cryostat for his studies.

REFERENCES

1. H. Walter, "Konstructionselemente aus der Kryotechnik", Verein Deutscher Ingenieure BW1954, (1998).
2. G.K. White, "The thermal and electrical conductivity of copper at low temperatures", Austr. J. Phys. 6, 397-404.
3. A.C. Rose-Innes, "Low Temperature Technique", The English Universities Press Ltd., London, 1974.
4. G. Klipping, Kaltetechnik 13, (1961) 250.
5. M/s Janis Research Company Inc-Website www.janis.com.
6. A.J. Croft, "Experimental Cryophysics", Ed. Hoare, F.H., Jackson L.C., Kurti, N., Butterworths, London 1961.
7. F. Din and Cockett A.H., "Low Temperature Techniques", Interscience, New York (1960).
8. D.H. Parkinson, "Experimental Cryophysics", Butterworths, London (1961).
9. McClintock, M., "Cryogenics", Reinhold Publishing Corp., New York, 1964.

7 CHAPTER

Thermometry

1. INTRODUCTION

Temperature is a measure of the degree of warmth or cold in a material. We can perceive through touch that one bath of water is warmer than another. To make such a concept precise we need to define a quantity called temperature in such a way that the higher the value of the temperature of a body, the hotter it is. We also have to devise practical and convenient methods of measuring temperature. The definition and methods must be such as to lead to the same value of the temperature of a body measured by different observers, using different methods.

A thermometer is a device to measure temperature. It uses a property of a material that changes as it is heated or cooled. This property can be the volume of the material, or the pressure of a gas at constant volume or some electrical property such as the resistance. We have to choose two different baths and arbitrarily set a value for the temperature T_1 for the cold bath and a value T_2, higher than T_1, for the temperature of the warmer bath. We bring the thermometric device in contact with the bath, wait till thermal equilibrium is achieved, and measure the value P of the property. Let $P(T_1)$ be the property value when the material is kept in the cold standard bath and $P(T_2)$ be its property value when the material is kept in the warm standard bath. If we now have a bath, the temperature T of which is to be measured, the thermometer is brought to thermal equilibrium with this bath and the property value P is measured. The temperature T of this bath is then defined by the relation

$$T = T_1 + \{[P - P(T_1)]/[P(T_2) - P(T_1)]\} \times (T_2 - T_1) \qquad (7.1.1)$$

The temperature of a bath defined in this way will depend both on the property that is measured and the material of the thermometer. For example, if we take the property to be the resistance of a metallic wire which forms the thermometer, there is no justifiable reason to expect that the way the resistance changes with temperature will be the same for wires of different material. This implies that temperature defined in this fashion will not be absolute in the sense that measurements made with different material and using different properties will generally give different values for the temperature of the same bath. This will create chaos.

To define an absolute temperature, we have to look for a device and a property such that the variation of the property with temperature will be independent of the working substance used. The device may only be a theoretical one that will not be possible to be realized in practice. Still, such an absolute temperature scale will be useful since it provides a benchmark against which the actual performance of practical thermometers can be tested.

2. ABSOLUTE TEMPERATURE

The Carnot engine is a heat engine in which the working substance undergoes a cycle of reversible changes in its thermodynamic state. In each cycle it absorbs a quantity, Q_1, of heat from a warm reservoir and delivers a smaller quantity, Q_2, of heat to a cold reservoir. The difference $W = (Q_1 - Q_2)$ is available as work. The important feature of the Carnot cycle is that its efficiency, η, defined by the relation

$$\eta = W/Q_1 = (1 - Q_2/Q_1) \tag{7.2.1}$$

is *independent* of the working substance in the heat engine or the nature of the property used in converting heat to work in the heat engine. This is a consequence of the second law of thermodynamics. So the efficiency of a Carnot engine can be used to define an *absolute temperature scale.* If we take the temperature of the standard cold reservoir as T_2, and the warm reservoir has a temperature T, we may define the efficiency by

$$\eta = (1 - T_2/T) \tag{7.2.2}$$

The coldest reservoir that one can use must have a temperature $T_2 = 0$ since the efficiency of any engine cannot be greater than 1. This is called the *Absolute Zero of Temperature* and all bodies must have a positive temperature on this scale. A scale of temperature defined thus is called the *absolute* or *Kelvin* scale.

2.1. Ideal Gas Scale

While it is nice to be able to define such an absolute scale, is there any practical way of realizing the Kelvin scale? Here the ideal gas laws come to our help. Boyle and Charles performed experiments on the pressure, P, of a given volume, v, of a gas in different warm baths. They found that if the gas is at low pressure, the product of the pressure and volume, Pv, remains constant in a given bath. This quantity increases as the bath becomes warmer. One can therefore take a mole of a gas at low pressure and determine the product PV (here V is the volume of 1 mole of the gas) in different baths. One can define a practical temperature scale using this product PV. If we take the temperature of a bath of melting ice as 0 and temperature of water boiling under atmospheric pressure as 100, we may define a gas scale of temperature by

$$T = \{[PV(T) - PV(0)]/[PV(100) - PV(0)]\} \times 100 \tag{7.2.3}$$

This relation implies that on this scale

$$PV = C + RT \tag{7.2.4}$$

The above constant, C, can be written as RT_0. In such a case,

$$PV = R(T + T_0) \tag{7.2.5}$$

Experiments show that (*a*) in the limit of low pressure all gases show the same behaviour and (*b*) the constants R and T_0 have the same values for all gases. The constant R has a value 8.316 Joules/K and T_0 has a value 273.15. We may now define a new scale of temperature, called the *Ideal Gas Temperature Scale*, in which the ice point temperature T_{2ig} = 273.15 and the temperature of boiling point water under atmospheric pressure T_{1ig} = 373.15. We may use any gas as our thermometric substance and measure the pressure at a constant volume of the gas. It is shown by experiments that the temperature of a bath deduced from such a measurement, in the limit of pressure of the gas tending to zero, is the same. Also it is independent of the gas. The relation

$$PV = RT_{ig} \tag{7.2.6}$$

is called the Charles law. At high pressures, the density of the gas becomes high and interatomic interactions play a role in causing deviations from Charles law.

2.2. Relation of Ideal Gas Scale to the Absolute Scale

How does the ideal gas scale of temperature relate to the absolute temperature scale based on Carnot cycle? If we use an ideal gas as the working substance in the Carnot engine and calculate from Charles law the heat, $Q_{1,}$ absorbed at the warm temperature T_{ig} and the heat, Q_2, rejected at the cold temperature T_{2ig} we find

$$Q_{2ig}/Q_{1ig} = T_{2ig}/T_{ig} \tag{7.2.7}$$

and hence,

$$\eta = (1 - T_{2ig}/T_{ig}) \tag{7.2.8}$$

Comparing the equations (7.2.2) and (7.2.8), we find that the ideal gas temperature T_{ig} should be proportional to the absolute temperature T_K. It is convenient to choose the constant of proportionality to be unity. Hence we see that the ideal gas temperature scale coincides with the absolute temperature scale. So the absolute temperature scale is realizable in practice using a gas thermometer filled at very low pressure (*i.e.* in the limit P tending to zero).

The ideal gas thermometer is called a *primary thermometer*. In actual practice a gas thermometer can be filled only at a finite pressure and there will be small deviations from Charles law. But one can apply corrections for temperatures measured by the gas thermometer and obtain the temperature on the absolute scale. These corrections will depend on the nature of the gas used.

In practice one cannot use the gas thermometer in experiments in any laboratory. One will have to use *secondary thermometers* that are small in size and weight and that are more convenient for measurement and control. There are a variety of such thermometers available. These are calibrated at a few fixed points using certain interpolation formulae as agreed upon by International convention. We shall describe the International Temperature Scale currently in use.

3. INTERNATIONAL TEMPERATURE SCALE (ITS 90)

The ITS 90 was adopted by the International Commission on Weights and Measures in 1989. This scale supercedes the practical temperature scale of 1968 and the provisional 0.5 to 30 K Temperature Scale of 1976. The temperature measured on the ITS 90 scale is written as T_{90}. The ITS 90 is defined in different temperature ranges as follows:

Table 7.1: Constants A_i (i = 0 to 9), *B* and *C* in Equation (7.3.1)

	^3He	^4He	
	0.65 to 3.2 K	*1.25 to 2.1768 K*	*2.1768 to 5 K*
A_0	1.053447	1.392408	3.146631
A_1	0.980106	0.527153	1.357655
A_2	0.676380	0.166756	0.413923
A_3	0.372692	0.050988	0.091159
A_4	0.151656	0.026514	0.016349
A_5	−0.002263	0.001975	0.001826
A_6	0.006596	−0.017976	−0.004325
A_7	0.088966	0.005409	−0.004973
A_8	−0.004770	0.013259	0
A_9	−0.054943	0	0
B	7.3	5.6	10.3
C	4.3	2.9	1.9

Between 0.65 to 5 K, T_{90} is defined in terms of vapour pressure of ^{3}He and ^{4}He. The temperature T_{90} is represented in terms of the vapour pressure by the following equation:

$$T_{90} = A_0 + \sum_{i=1}^{9} A_i\{[\ln(P) - B]/C\}^i \tag{7.3.1}$$

The vapour pressure *P* is in Pascals. Table 7.1 gives the constants for ^{3}He and ^{4}He and the temperature range in which the constants are valid is indicated.

Between 4.2 K to triple point of Ne, a ^{4}He gas thermometer calibrated at three fixed points (at the triple point of Ne, triple point of e-Hydrogen, and one other temperature between 3 and 5 K as determined by vapour pressure thermometer) is used. The equation to be used to obtain T_{90} is

$$T_{90} = a + bP + cP^2 \tag{7.3.2}$$

Here P is the pressure of the Helium gas. The constants a, b and c are determined from the calibration at the above three temperatures.

Between the triple point of e-hydrogen and melting point of silver, a platinum resistance thermometer calibrated at specified sets of defined fixed points using specific interpolations defines T_{90}. We define $W(T_{90})$ by the ratio $R(T_{90})/R(273.16)$ where $R(T)$ is the resistance of the Pt thermometer at temperature T. The thermometer should be made of pure strain-free Platinum wire satisfying at least one of the following two conditions:

$$W(MP \text{ of Ga}) \geq 1.11807 \tag{7.3.3}$$

and

$$W(TP \text{ of Hg}) \geq 0.844235 \tag{7.3.4}$$

If it is to be used till the freezing point of Ag it must also satisfy the condition

$$W(FP \text{ of Ag}) \geq 4.2844 \tag{7.3.5}$$

From 13.8033 to 273.16 K the temperature can be calculated from $W(T)$ by the relation

$$T_{90} = B_0 + \sum_{i=1}^{15} B_i[\{W^{1/6}(T) - 0.65\}/0.35]^i \tag{7.3.6}$$

The constants B_i are listed in Table 7.2. Table 7.3 gives the temperature of a few fixed points.

Table 7.2: Constants for Pt resistance thermometer in the relation (7.3.6)

Constant	*Value*	*Constant*	*Value*
B_0	0.183324722	B_8	–0.075291522
B_1	0.240975303	B_9	–0.056470670
B_2	0.209108771	B_{10}	0.076201285
B_3	0.190439972	B_{11}	0.123893204
B_4	0.142648498	B_{12}	–0.029201193
B_5	0.077993465	B_{13}	–0.091173542
B_6	0.012475611	B_{14}	0.001317696
B_7	–0.032267127	B_{15}	0.026025526

Table 7.3: Fixed Point Temperatures; e-Equilibrium; TP-Triple Point; MP-Melting Point; FP-Freezing Point

Fixed Point	*Temperature (K)*	*Fixed Point*	*Temperature (K)*
e-Hydrogen TP	13.8033	Gallium MP	302.9416
Neon TP	24.5561	Indium FP	429.7485
Oxygen TP	54.3584	Tin FP	505.078
Argon TP	83.8058	Zinc FP	692.677
Mercury TP	234.3156	Aluminium FP	933.473
Water TP	273.16	Silver FP	1234.93

4. SECONDARY THERMOMETERS

A secondary thermometer is a device of low thermal capacity having a property, which changes with temperature. It should have a well-characterized and reproducible response to variations in temperature. It will be useful if the response can be fitted to a well-defined equation containing a few parameters that can be obtained by calibrating the thermometer at a few fixed points. The sensitivity of the thermometer (*i.e.* the change in signal for one degree change in temperature) should be high. At the same time the thermometer should respond quickly to changes in temperature. This is achieved by making the thermal capacity of the thermometer small and having a good thermal contact with the object, the temperature of which is to be measured.

For cryogenic thermometry it is necessary to have long leads connecting the sensor at low temperature to electrical leads at room temperature. One must take adequate precautions to see that the heat leaking through the leads is small. Otherwise the temperature indicated by the thermometer will be higher than the actual temperature of the object. One should use as thin and as long leads as possible to increase the thermal resistance of the leads. It will be good to anchor the leads to a point that is at the temperature of the object at low temperature. Then all the heat from room temperature will go to the cold object and ultimately to the refrigerating medium used to cool the object. Heat leak to the thermometer will become small. This is indicated in Figure 7.1.

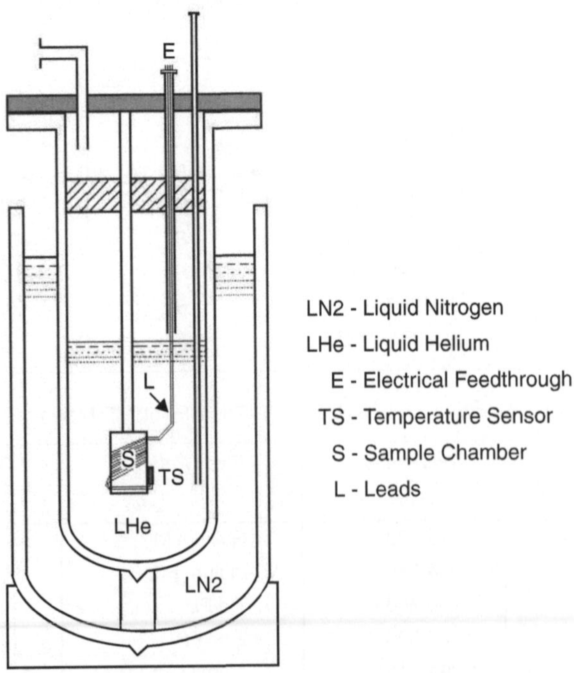

Figure 7.1: Schematic for showing the anchoring of thermometer leads

The thermometer leads from the feed through (*E*) at the room temperature flange is wrapped round the specimen chamber and fixed to it with a thin layer of *GE* varnish before being connected to the thermometer. Such *anchoring* is necessary in all thermometric measurements at low temperature.

5. RESISTANCE THERMOMETERS

One of the most widely used techniques for measurement of temperature is resistance thermometry. In this technique the resistance of a material is measured and the temperature is deduced from the value of resistance. One can have materials with a *positive temperature coefficient* (PTC) of resistance. In these materials the resistance increases with increase in temperature. This is usually the behaviour of metals and metallic alloys. The semi-conducting materials, on the other hand, have a *negative temperature coefficient* (NTC) of resistance. In these materials, the resistance increases (often by a few orders of magnitude) as the sensor is cooled. In the case of PTC sensors one uses a constant current source to drive a constant current through the material. The voltage across the sensor is measured with a digital voltmeter. As the temperature falls the power dissipated in the sensor will decrease. With NTC materials, it is usual to use a constant voltage across the resistor and measure the current. The power dissipated is inversely proportional to the resistance. So, with this method of measurement, the power dissipated will decrease in NTC sensors as the sensor cools.

5.1. Self Heating

Why are we worried about the power dissipated in the sensor? This power will have to be conducted away to the cold object, the temperature of which is being measured. If the thermal resistance of the contact between the thermometer and the object is $\rho_{contact}$ and P is the power dissipated in the sensor during measurement, there will be a temperature difference ΔT between the thermometer and the object given by

$$\Delta T = \rho_{contact}\, P \tag{7.5.1}$$

If the object temperature is T, the temperature recorded by the thermometer will be $T + \Delta T$. This is called the self-heating effect. The contact resistance increases as the temperature falls and varies as T^{-3} below 1 K. It is therefore necessary to keep P small enough so that ΔT is a small fraction of T. The lower the temperature the smaller should be the value of P. That is the reason for using a constant current mode of measurement with PTC sensors and a constant voltage mode of measurement with NTC sensors. The power dissipated depends on the square of the current while the measuring signal will be proportional to the current. It is therefore possible to choose a current value that will give a large enough signal to measure the temperature accurately and that, at the same time, will keep the power dissipated small enough to make the error ΔT smaller than a pre-assigned value.

5.2. Two Wire and Four Wire Measurement

Figures 7.2(*a*) and 7.2(*b*) show the connections for a two wire and a four wire measurement of the resistance of the sensor. In the two wire measurement the DMM measures the voltage across the resistance thermometer and the current leads. So the resistance measured includes the resistance of the current leads, which will vary with temperature in a manner different from the behaviour of the resistance thermometer. Unless the resistance of the current leads is negligibly small compared to the resistance of the thermometer, the measured temperature will be in error.

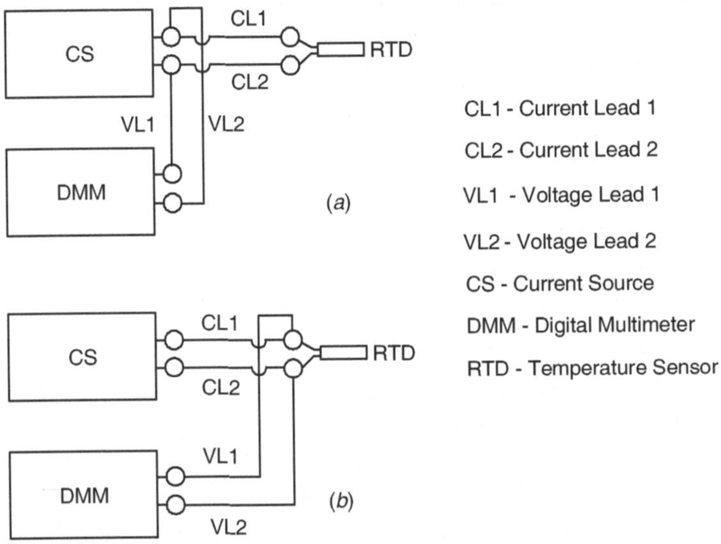

Figure 7.2: (*a*) Two wire measurement, (*b*) Four wire measurement

In the four wire measurement the current and voltage leads are connected to the terminals of RTD. So the DMM measures only the voltage across RTD and the resistance of the current leads does not enter into the measurement.

5.3. AC Measurement

One can also measure the resistance of RT by AC methods using either a bridge circuit or a lock in amplifier (see Chapter 10). This involves expensive instrumentation and is justified only if one requires a very high sensitivity in measurement.

5.4. Platinum Resistance Thermometer (PRT)

The platinum resistance thermometer is widely used in the temperature range 20 K to room temperature and above. The thermometer is made of a thin, high purity wire of platinum mounted strain-free and encapsulated in a ceramic tube or a glass tube. It comes with two leads. The PRT has a diameter of about 2 mm and a length varying between 5 to 20 mm. For four wire measurement the voltage and current leads will have to be soldered to the two leads of the PRT. The resistance thermometer usually has a nominal resistance of 100 Ohms at 273.15 K. For such a thermometer the sensitivity is roughly about 0.4 Ohm/K from 40 K to 500 K. Below 30 K the sensitivity falls off steeply. The temperature variation of resistance is shown in Figure 7.3.

If the thermometer is calibrated, its accuracy in measuring temperature is around 5 to 10 mK and reproducibility is also of the same magnitude. The response time of the thermometer depends on (*a*) its mass and (*b*) the temperature. The response time is a few seconds at low temperature and increases as the temperature increases. The excitation current is usually 1 mA, which will give a self-heating power of 100 microwatts at 273 K.

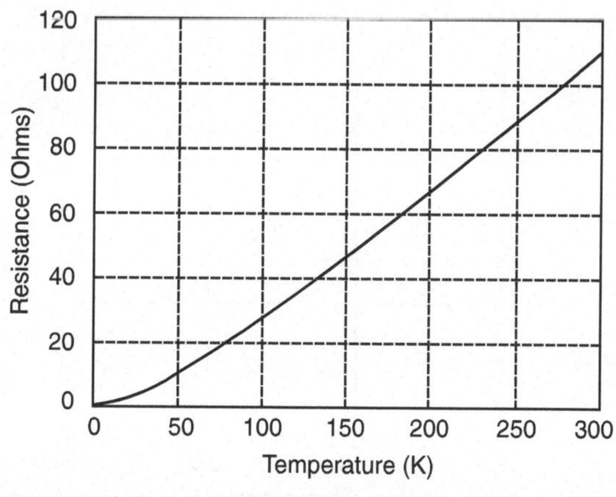

Figure 7.3: Temperature variation of resistance of a 100 Ohm Platinum RT (from www.tc.co.uk)

The Platinum thermometer can be used above 50 K in magnetic fields as the magneto-resistance of platinum causes only an error of a few per cent in the measurement of temperature.

5.5. Rhodium-Iron Resistance Thermometer

The temperature variation of resistance of Rhodium-iron alloy depends on its composition. It is possible to choose a composition for which the resistance increases monotonically in the range 1.4 to 300 K as the temperature increases. Thermometers are made either with a thin film of Rh-Fe alloy or with a thin wire. Because of the small thermal capacity in the former case, the response time is much shorter (of the order of a few milliseconds at 4.2 K to about twenty milliseconds at 300 K) than the response time (a few seconds) with the wire wound thermometer. The sensitivity is high from 100-300 K. Below 100 K the sensitivity drops till 30 K. The sensitivity rises again to a high value below 10 K. This thermometer is unsuitable for measurement in magnetic fields below 77 K.

5.6. NTC Resistance Thermometers

5.6.1. *Doped Germanium Resistance Thermometer*

Germanium (Ge) is a semi-conductor and appropriate doping can alter its electrical conductivity. The resistance increases sharply as the temperature is reduced. Various models are available with different suppliers, for use in the temperature range 0.05 K to 30 K. The rapid increase in sensitivity below 4.2 K makes it well suited for sub-milli Kelvin control of temperature. A typical resistance temperature curve for a GRT is shown in Figure 7.4.

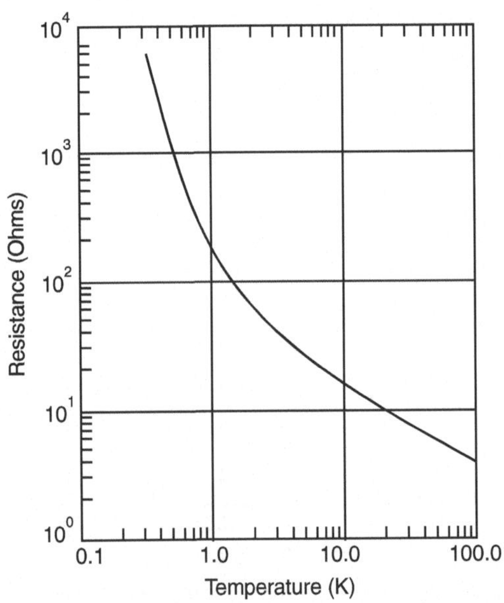

Figure 7.4: Typical variation of the resistance of a Germanium Thermometer with temperature (from www.lakeshore.com)

It is seen that above 30 K the sensitivity of the thermometer becomes poor. It has to be used with a constant voltage excitation. This excitation voltage is chosen so that the self-heating power at 4.2 K is about 0.1 μW and at 0.5 K is about 0.1 pico-Watt. Typical thermal response time is a few hundred milliseconds at 4.2 K and a few seconds at 77 K.

5.6.2. *Cernox Thermometers*

These are ceramic oxides with a negative temperature coefficient of resistance. They are useful in the temperature range 1.4 to 300 K. These thermometers are characterised by a very low magneto-resistance and so are useful to measure temperature in the presence of high magnetic fields. These are used with constant voltage excitation of a few millivolts such that the self-heating power is about 10 μW at 300 K and 0.1 μW at 4.2 K. The thermal response time is a few milliseconds in the range 0.3 to 4.2 K. Figure 7.5 shows the resistance versus temperature graph of commercial Cernox thermometers.

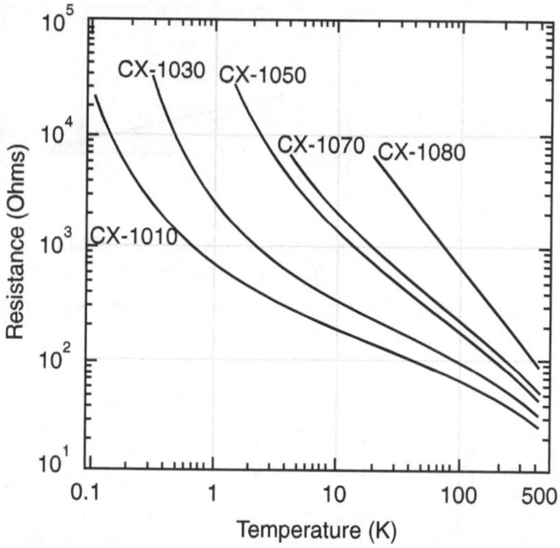

Figure 7.5: Resistance vs temperature of commercial Cernox Thermometers (from www.scientificinstruments.com)

5.6.3. *Other NTC Resistance Thermometers*

Carbon glass sensors, which consist of carbon fibres trapped in a glass matrix, can be used from 1.4 to 300 K. They have a negative temperature coefficient of resistance. Below 10 K they have a high sensitivity. However above 100 K their sensitivity falls below 1 Ohm/K. Above 100 K they can only provide rough temperature measurement to an accuracy of 0.1 K. They have a low magneto-resistance, this being negative below 20 K. They can be used in magnetic fields as high as 19T and the correction for the temperature amounts only to a few per cent.

Ruthenium oxide sensors are thick film thermometers. They are useful in the range 50 mK to 30 K. The magnetic field dependence of resistance is small.

Allen-Bradley carbon resistors have been used for thermometry. They are inexpensive. But they need frequent calibration.

6. DIODE THERMOMETERS

At a constant current, the voltage across a forward biased *p-n* junction diode changes with temperature. This property is used in diode thermometry. The advantage of diode thermometers is their sensitivity (change in forward bias voltage for unit change in temperature) is large and nearly constant from 300 K down to about

25 K in the case of Si diodes and 40 K in the case of GaAs diodes. Below 25 K in Si diodes and 40 K in GaAs diodes, the forward voltage increases rapidly as the temperature decreases.

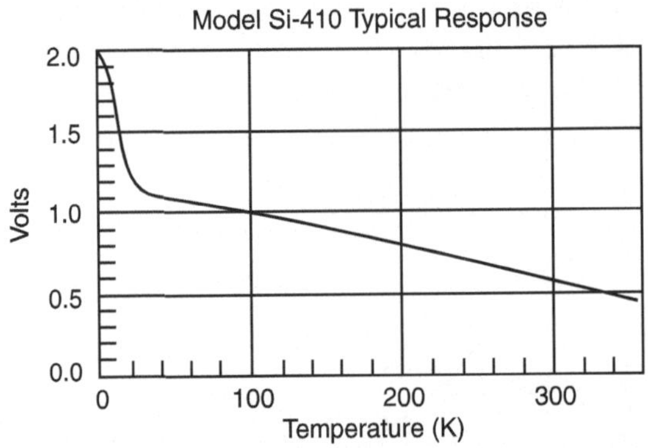

Figure 7.6(*a*): Forward voltage of a Si diode as a function of temperature at constant current (from www.scientificinstruments.com)

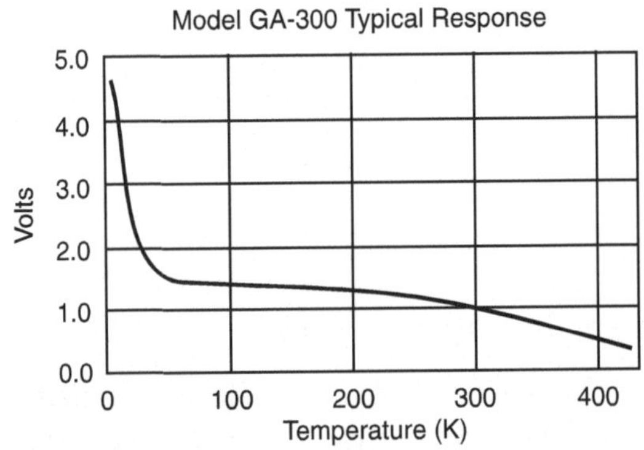

Figure 7.6(*b*): Forward voltage of a Ga-As diode as a function of temperature at constant current (from www.scientificinstruments.com)

The forward voltage as a function of temperature at a constant current is shown in Figure 7.6 for both Si and GaAs diodes. Si diodes are relatively inexpensive. Though they are less sensitive than GaAs diodes they are more stable. Above 1 K one can achieve an accuracy of about ±25 mK in temperature measurement. The excitation current is 10 μA. Problems can arise if there is an AC component of current from the DC source. One should take precautions like proper electrical shielding and grounding to get precise measurements. The GaAs diodes can be used in moderate magnetic fields.

7. THERMOCOUPLES

Thermocouples have been used often in the past for the measurement of cryogenic temperatures. They are inexpensive. When two dissimilar metals *A* and *B* are joined together and the two junctions are maintained at different temperatures T_1 and T_2 an electromotive force is generated which can be measured on a digital multimeter. The typical measurement arrangement is shown in Figure 7.7.

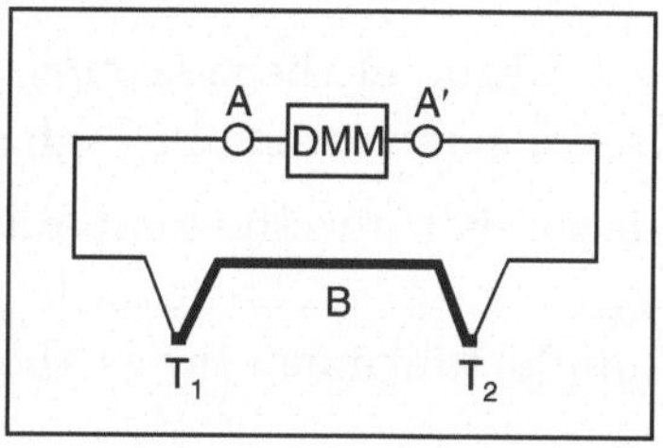

Figure 7.7: *A* and *B* are wires of two dissimilar metals joined together to form two junctions maintained at temperatures T_1 and T_2. The open ends of *A* are connected to the terminals of a DMM.

If T_1 is different from T_2 the DMM indicates an emf in micro- or milli-volts. This is called the thermo-emf. The thermo-emf is a function of temperature difference $(T_1 - T_2)$ provided the junctions of leads *A* and *A*' to the terminals of the DMM are at the same temperature. If T_2 is the temperature of a reference bath such as melting ice then one can measure the unknown temperature T_1. Usually the reference bath is dispensed with. The open ends of *A* and *B* are connected to the DMM. But at the junction to the DMM a resistance thermometer measures the temperature and produces an appropriate correction voltage by a bridge circuit.

There are three standard thermocouples used in the range of cryogenic temperature. These are type *E* (chromel-constantan), type *K* (chromel-alumel) and type *T* (copper-constantan). In Figure 7.8 the thermoelectric emf of these standard junctions are plotted in millivolts against temperature T_1, taking the reference temperature to be that of melting ice (273.15 K). The slope of the curve gives the sensitivity. We see that the curves flatten out at the low temperature end. The sensitivity falls at low temperature and so these thermocouples are not usable below 30 K. For example the thermoelectric power at 20 K for *E*, *K* and *T* type thermocouples are 8.5, 4.1, and 4.5 μV/K.

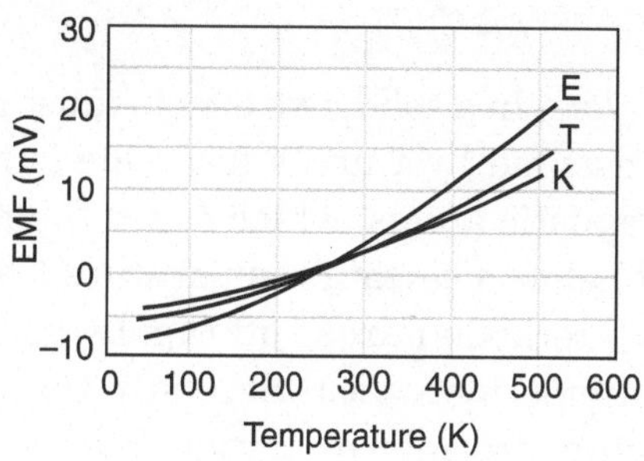

Figure 7.8: Thermo-emf of standard thermocouples with the reference junction at 273.15 (from www.scientificinstruments.com)

Since the wires find themselves in a region where temperature gradient exists, it is essential that the wires *A* and *B* must be homogeneous. Otherwise the temperature gradient along the length of the thermocouple wires will produce an emf of its own.

The Gold- (0.7 at %) Fe – Chromel thermocouple is suited for temperatures below 30 K as its thermo-electric power remains high at 15 μV/K above 10 K.

Using thermocouples one can measure the temperature to an accuracy of about 1 K. Relative changes in temperature can however be measured more accurately. The advantage of thermocouple thermometry is the low response time of the thermometer.

8. CAPACITANCE THERMOMETERS

The dielectric constant of Strontium Titanate shows an appreciable variation with temperature. A capacitance of a few nano-farads made with this material as the dielectric medium can be used as a thermometer below 100 K. The capacitance will have to be measured by an AC bridge with about a few volts of excitation. The disadvantage of these thermometers is that the calibration changes after thermal cycling. They have to be re-calibrated frequently. The advantage of the capacitance thermometer is that it is totally insensitive to magnetic fields. The capacitance thermometer is useful for controlling temperature in the presence of magnetic fields. Its thermal response time is high, of the order of minutes.

9. MAGNETIC SUSCEPTIBILITY THERMOMETER

Temperatures of a few tens of a milli-Kelvin can be obtained either by adiabatic demagnetization of a paramagnetic salt or with the help of the dilution refrigerator. For the measurement of such low temperatures the magnetic susceptibility thermometer can be used. The paramagnetic susceptibility of a dilute system of magnetic atoms varies with temperature as

$$\chi = C/T \tag{7.9.1}$$

This is known as the *Curie law*. In a solid, two effects cause a change in the behaviour of the susceptibility represented by Curie's law. One is the exchange interaction between neighbouring magnetic atoms, which cause the magnetic moments of the atoms to become ordered below a certain temperature leading to ferro-, antiferro- or ferri-magnetism. This interaction drops off rapidly as the distance between the neighbouring magnetic atoms increases. Secondly the degenerate levels of the ground state of the free atom are split by crystal field effect.

The paramagnetic salt should be in good thermal contact with the object the temperature of which is to be measured. The magnetic susceptibility is measured by using a Hartshorne bridge or a ratio transformer system. Cerium magnesium nitrate is the thermometric substance commonly used below 1 K since its magnetic ordering temperature is 4 mK. The other material is Pd containing a few parts per million of iron atoms. The spin freezing temperature of this material is 0.1 mK and it has a giant atomic magnetic moment of 10 Bohr magnetons.

10. VAPOUR PRESSURE THERMOMETRY

The pressure exerted by the saturated vapour over the liquid surface is a definite function of temperature and hence can be used to measure the temperature of the liquid. In fact, in the International Scale of Temperature (ITS 90), the vapour pressure of ^{3}He and ^{4}He are used for defining the temperature scale between 0.65 K and 5 K. With a good pressure measuring arrangement, the vapour pressure thermometer is an excellent secondary standard, since the temperature response depends on the physical property of a pure element. The vapour pressure is expressed by semi-empirical formulae with fit coefficients. The vapour pressure equations for many common cryogenic fluids are given in Table 7.4 and the constants in Table 7.5.

Table 7.4: Vapour pressure equations for some cryogenic fluids (from Reference (2))

Substance	*Equation*	*Units*	*Scale*
Oxygen	$\ln P = \sum_{n=1}^{8} A_n T^{n-1}$	P atm T K	NBS 55 NBS 48
Nitrogen	$\ln P = A_1/T + A_2 + A_3 T + A_4 (T_c - T)^{1.95}$ $+ A_5 T^3 + A_6 T^4 + A_7 T^5 + A_8 T^6 + A_9 \ln T$	P atm T K	NBS 55
Neon	$\ln P = A_1 + A_2/T + A_3 + A_3 T + A_4 T^2$	P torr T K	NBS 55
Hydrogen	(1) $13.803 < T < 21$ K $\ln P = A_1 + A_2/T + A_3 T$ (2) $20.268 < T < 29$ K $\ln P = A_4 + A_5/(T + A_6) + A_7 T$ (3) $29 < T < 32.976$ K $P = P_0 + A_8 (T - 29)^3 + A_9 (T - 29)^5 + A_{10}(T - 29)^7$	P atm T K ,, ,,	NBS 55 ,, ,,
Helium	$\ln P = \sum_{j=1}^{10} A_j T^{(2-j)}$	P μm Hg T K	$T = T_{58} + 0.001$ $+ 0.0002 T_{58}$

Table 7.5: Values of the coefficients in the vapour pressure equations in Table 7.4 (from Reference (2))

Coefficient	*Oxygen*	*Nitrogen*	*Neon*	*Hydrogen*	*Helium*
A_1	–62.5967185	0.8394409444×10^4	7.46116	1.772454	–3.9394635287
A_2	2.47450429	$-0.1990045259 \times 10^4$	–106.090	-4.436888×10	1.4127497598×10^5
A_3	$-4.665973315 \times 10^{-2}$	$-0.7252229165 \times 10^5$	-3.56616×10^{-2}	2.055468×10^{-1}	$-1.6407741565 \times 10^3$
A_4	$5.48202337 \times 10^{-4}$	0.1022850966×10^5	4.11092×10^{-4}	2.000520	1.1974557102×10^4
A_5	$-4.09349868 \times 10^{-6}$	$0.5560632825 \times 10^{-3}$		-5.009708×10	$-5.5283309818 \times 10^4$
A_6	$1.91471914 \times 10^{-8}$	$-0.5944544662 \times 10^{-5}$		1.0044	1.6621956504×10^4
A_7	$-5.13113688 \times 10^{-12}$	$0.2715433932 \times 10^{-10}$		1.7848495×10^2	-3.25212828×10^4
A_8	$6.02656934 \times 10^{-14}$	$-0.4879535904 \times 10^{-10}$		1.317×10^{-7}	3.9884322750×10^4
A_9		0.509536824×10^3		-5.926×10^{-5}	$-2.7771806992 \times 10^4$
A_{10}				3.913×10^{-4}	8.3395204183×10^4

The different fixed points of common vapour pressure thermometers are given in Table 7.6.

Table 7.6: Fixed points for common vapour pressure thermometer fluids (from Reference (2))

Fluid	*Triple Point (K)*	*Normal Boiling Point (K)*	*Critical Point (K)*
Oxygen	54.351	90.18	154.576
Nitrogen	63.148	77.347	126.20
Neon	25.54	27.09	44.40
Hydrogen	13.803	20.268	32.976
Helium-4	2.177 (λP)	4.224	5.2014

The vapour pressure thermometer is useful for temperature measurement in the range from the triple point to the critical point of the liquid. One of the important advantages of vapour pressure thermometer is the extreme sensitivity in the range over which it can be used. The disadvantage is that it is useful only over a limited temperature range. The thermometers are most accurate in the range of the normal boiling point of the liquid chosen. The constants for the vapour pressure equations are given in Table 7.5 and the fixed points in Table 7.6. The data given in the above tables may require slight modification to suit the ITS 90 scale.

The apparatus for vapour pressure determination is quite simple. There are many variations of vapour pressure thermometers, but one described by Scott is shown in Figure 7.9. The copper bulb having the liquid is connected to an accurate

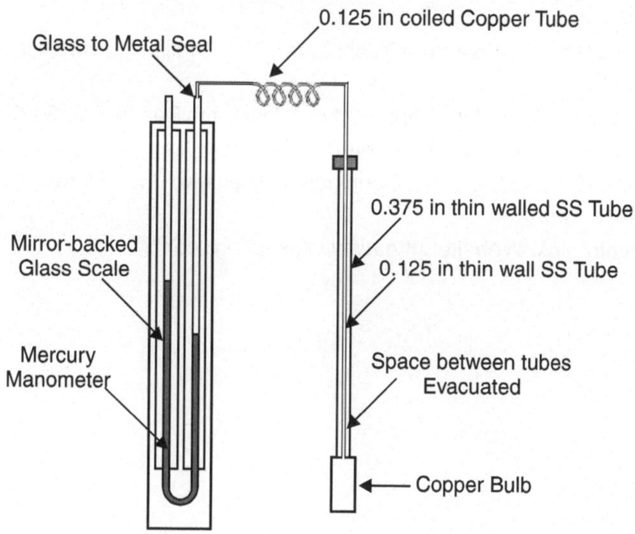

Figure 7.9: A set-up for vapour pressure thermometry (after Scott)

mercury manometer through a thin walled vacuum jacketed stainless steel tube and a coiled copper tubing. It should be obvious that the tubing between the bulb and manometer should nowhere be colder than the bulb since this can cause refluxing and the pressure readings will be low.

Also, the gas purity is quite important. With Hydrogen, the ortho/para ratio is an important parameter due to the strong dependence of the vapour pressure on this composition. In the case of helium, hydrogen is the possible source of impurity, since all the other impurities will be solidified. As mentioned earlier, the vapour pressure thermometry is quite sensitive. It also has good response and is not affected by magnetic fields and does not need calibration.

11. CONCLUSION

There are other thermometric techniques like noise thermometry and nuclear orientation thermometry, which are outside the purview of the book. Since temperature measurement is an important component of any low temperature experiment one must pay special attention to thermometry.

REFERENCES

1. Thomas M. Flynn, "Cryogenic Engineering", Marcel Dekker Inc., N.Y., 1997.
2. Randall F. Barron, "Cryogenic Systems", 2nd edition, Oxford University Press, New York, 1985.
3. Scott R.B., "Cryogenic Engineering", D. Van Nostrand Co., Inc., Princeton, New Jersey, 1959.
4. McClintock M., "Cryogenics", Reinhold Publishing Corp., New York, 1964.
5. Guy K. White, "Experimental Techniques in low temperature Physics", 3rd edition, Oxford University Press, New York, 1979.
6. A.C. Rose-Innes, "Low Temperature Technique", The English Universities Press Ltd., London, 1974.
7. Scientific Instruments, 440 West Tiffany Drive, West Palm Beach, FL 33407, USA, Product data sheets, website: http://www.scientificinstruments.com.
8. Lab Facility, Unit 3B, Outgang Lane, Dennington, Sheffield, U.K., Product data sheets. Website: http://www.labfacility.co.uk.
9. Lakeshore Cryotronics, Website: http://www.lakeshore.com.

PART II TECHNIQUES FOR THE MEASUREMENT OF PHYSICAL PROPERTIES OF SOLIDS AT LOW TEMPERATURES

8

CHAPTER

Basic Concepts of Measurements

1. INTRODUCTION

In this chapter we will discuss some of the basic concepts that are needed for doing solid-state measurements at low temperatures. The focus of this chapter is to learn about some basic concepts that are needed in all solid-state measurements rather than on details specific to a given technique. These concepts of measurements are the fundamentals that can be innovatively utilized in doing different measurements. Often solid-state measurements are done using commercial set-ups which are available as a full system with both soft and hard wares. These set-ups can do a class of measurements rather routinely. The users of such equipments would find the chapter (and subsequent chapters useful) because it will explain the basics of some of these measurements and one need not use the "black-box" approach for using these set-ups. However, the book is focussed on experimenters who would like to design and fabricate their own experiments and also would like to do experiments that are not available in the commercial-set-ups. In any case it pays to be innovative for an experimenter. The main approach used in this book here is of "system integration" where basic measuring instruments like Lock-in amplifiers, Digital voltmeter and ammeter, Electrometer, Synthesizers, Frequency Counters and Digital Scopes etc. are connected as a system that is automated through a computer.

All measurements use a physical law or a combination of laws. Deriving a "number" or a value for the physical quantity is at the heart of measurements. There cannot be a physical measurement without quantification. And no physical theory can be complete without the measurements. Briefly, the underlying theory or the law and the quality of the measurements would decide the output and also the reliability of an experiment. As an experimenter it is of primary importance that one knows the capability of the measurement set-up used. For instance, to verify experimentally a theory one may need to measure the change in resistance to say 1 part in 10^6 (1 ppm). If the apparatus gives a precision of 1 part in 10^4, it will be futile to do the experiment. Similarly if 1% accuracy is all that is needed for an experiment, doing the experiment to 1 ppm accuracy is a waste of time and resources. The knowledge of the experiment, the physics that one is looking for and of course the skill of the experimenter will finally decide the success. In the following we start by defining the structure of a physical experiment.

In Figure 8.1 we explain the basic physical measurements with the help of a cartoon. It is the sample *environment* that fixes certain physical parameters and typically the experimenter may vary one or more parameters (which we call the variables) so that we can study the quantity to be measured as a function of the variables. The physical quantity to be measured is obtained as a response function where we measure the response of the system when a particular excitation is given to it.

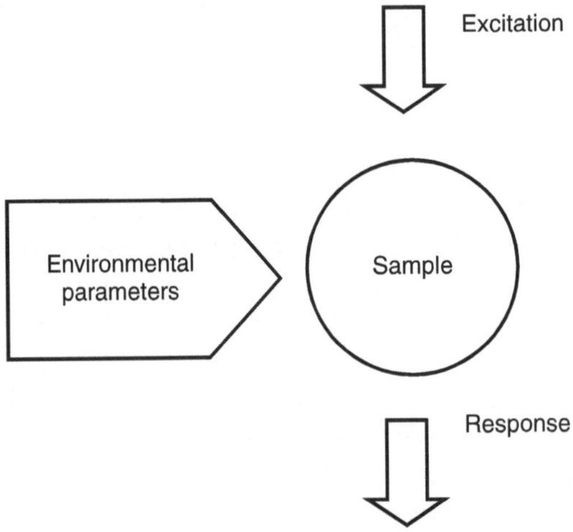

Figure 8.1: Basic physical measurement on a sample in an environment

We explain the above cartoon in the context of measuring the magneto-resistance (MR) of a sample. MR is the change in the resistance (R) of the sample in a magnetic field (H). The change in resistance $\Delta R(\equiv R(H) - R)$ may depend on H as well as on the measurement temperature. In this case generally the *excitation* is a fixed current (or an alternating current) and the *response* is a dc voltage (or an ac voltage). When the field is kept fixed, as in Figure 8.2, the MR is measured as a function of temperature. This is measured by measuring the resistance R in two magnetic fields $0T$ and a finite field ($6T$). The MR is measured from the change in R when H is applied. In this case the *temperature is the variable and H is the fixed environmental parameter.*

One can also take the data by keeping the temperature fixed and by varying the magnetic field. *In this case the temperature is the fixed environmental parameter and H is the variable.* For most physical measurements the environmental parameters are generally Temperature (T), Magnetic field (H), Electric field (E), Hydrostatic pressure (P), Uni- or bi-axial Strain, and Intensity of charged or electromagnetic excitation. What environmental parameters to use and which one should be kept constant and which one as variable depend solely on what information we would like to extract from the measurements.

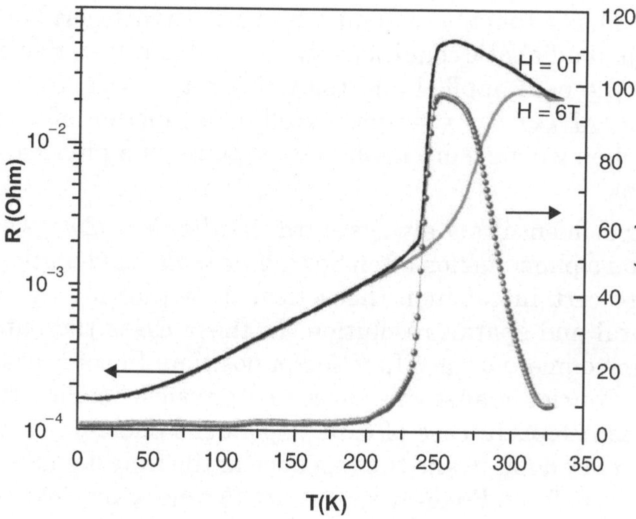

Figure 8.2: Example of Magneto-resistance (MR) of a solid as a function of temperature *T*. In this case the MR is obtained by keeping the applied field fixed and measuring the resistance as the temperature is varied. The MR is defined as MR = {[*R*(*H*)-*R*]/*R*}. MR is negative since the resistance decreases in an applied field.

How effective will be the change of a physical parameter? This is an important question that will help one to assess the physics underlying the phenomenon that is being investigated. In fact success of an experiment depends on the how best one can use a set of parameters, and vary them in a controlled way to extract information on the physical phenomena. To see the effect of variation of a parameter one may start from some ground rules. First ground rule is that the physical parameter that is changed must couple to the physics of the phenomenon. This can be determined empirically in the absence of a theory or a model. Often this can be based on a hypothesis. But a theory or a model is most helpful when deciding these factors.

Another ground rule is how much change in the energy such a variation of parameters can bring about. For example a magnetic field of $1T$ is expected to bring about a change in the energy in the scale of $\approx \mu_B H$, where μ_B is the Bohr magneton. As a result one would expect a noticeable change in physical properties at temperatures $T \approx \mu_B H/k_B = 0.7$ K, where k_B is the Boltzmann constant. There are of course noticeable exceptions where a small magnetic field can bring about a large change at high temperature. We can rationalize this in the following way. Suppose there are two phases A and B with free energies F_A and F_B respectively. The free energies themselves can be large (>> $\mu_B H$) but the free energy difference $\Delta F \equiv F_A - F_B \approx \mu_B H$. Then even a small magnetic field can tilt the balance and this will lead to a phase transition.

In measurements of some physical quantities the response itself can be a part of the environmental parameter. A nice example of this is measurement of heat capacity C of a solid as a function of temperature T. In this case heat ΔH is applied to a material (kept at adiabatic condition) and the temperature rise ΔT is measured. The excitation is the heat applied and the temperature rise is the response. The heat capacity $C \equiv (\Delta H/\Delta T)$ is then measured as a function of T itself. A similar situation arises when we measure nonlinear response of a physical quantity like a dielectric constant.

A physical experiment is not always static, neither is it always space-averaged. The physics behind a phenomenon often has a *time scale* and length scale associated with it. There are certain common theoretical tools that are used to handle the data with temporal and spatial resolution. In these cases the data can be taken directly in the time domain or as a function of position. In some cases the data can be taken in the Fourier transform space. In physics Fourier transform is an extremely important tool. In case of time-dependent phenomena the data can be taken in the frequency domain and the response in the time domain can be obtained from the Fourier transform. Roughly speaking the time scale τ of a physical process will show up in the frequency domain at frequency $f \approx \frac{1}{2\pi\tau}$ and the *length scale* of λ will show up in the scattering experiments (e.g., neutron, X-ray or optical) that probe the reciprocal space at a wave vector $\approx \frac{1}{2\pi\lambda}$. In modern day experimentation much attention is paid to acquire spatially and temporally resolved data and a number of advanced experimental techniques are used for that. In this book we will not address these issues specifically.

2. SPATIALLY AND TEMPORALLY RESOLVED EXPERIMENTS

We have mentioned before the need for spatially and temporally resolved data that can be obtained either from direct space or time domain experiment or from reciprocal space or frequency domain experiments. In this section we discuss briefly the basic physics contents taking some examples.

The most common experiment that takes data in reciprocal space and gives information on the structure in real space is X-ray, neutron scattering, electron or even optical scattering. We refer the readers to standard text books to learn more about the physics of these scattering techniques [1]. The basic concept involved in these experiments is the structure factor. This is at the core of a large number of spatially resolved experiments. A similar concept in time domain experiment is the correlation function.

However, there are a number of direct imaging techniques that offer high degree of spatial resolution without taking recourse to structure factor. These involve Scanning Electron Microscope (SEM) and the family of Scanning Probe Microscopes (SPM). SPM is a rather powerful tool that can give high-resolution 3D images and one can obtain atomic resolution. SPM can also be very easily adapted to Ultra High Vacuum (UHV) environment, magnetic field and also low and high temperature. This has become an indispensable tool for some low temperature investigations. Modern day SEM can achieve resolution down to 3.5 nm with Field Emission Guns (FEG). The resolution can be improved to 1.8 nm. The environmental scanning SEM can work even in a very low vacuum and can handle wide variety of substances. In addition, it can provide useful spatially resolved elemental analysis through the characteristic X-rays that comes out when the electron hits the sample.

Transmission Electron Microscope (TEM), which depends on electron diffraction, has reached a very advanced stage to provide spatially resolved structural and chemical information down to nanometer scales. Through high-resolution lattice imaging they can give a resolution down to atomic scale.

While scattering experiments, and imaging by probes like Scanning Probe Microscopes and Electron microscopes can give us spatially resolved information, there are certain other ways to obtain information on length scales. Interference experiments, interestingly, can be very good probes of length scales. The key concept here is of *phase coherence*. Any physical parameter that can break the phase coherence can be used to probe the *length scale* of phase coherence. We take one example from the field of low temperature physics. Any inelastic process breaks the time reversal symmetry and thus can break the phase coherence. In this context, temperature can be a good probe of length scale! In the field of low temperature physics this becomes very relevant. This has been the basic concept in low temperature studies on the phenomenon called electron localization [2]. (Note: In disordered solids, when the disorder is too large and the elastic mean free path is too small one observes a phenomenon called electron localization. This is quantum interference of electrons, which are back scattered at the defects. This phenomenon can give rise to high electrical resistance of a solid.) Similarly a static magnetic field, which introduces an additional phase on the orbital motion of electron, can provide a good measure of the phase coherence length of electron. In such a case when a magnetic field is applied it can break the time reversal symmetry that is crucial to get electron interference. The weakening of the localization due to the magnetic field leads to a lowering of the electron resistance, a phenomena known as negative magneto-resistance. In this case the phase coherence length of the electron can be obtained from the magnetic field.

3. THE BASIC BUILDING BLOCKS AND AUTOMATING THE EXPERIMENT

In Figure 8.1 we described the four basic components of an experiment-source (excitation), detector, the environment and the system on which the experiment is carried out, which we call the sample. The basic detectors in modern measurements convert the data into electrical domain through transducers that allow digitization and subsequent signal processing. All the modern day measurements are done though digital equipments and use computers. In fact experiments should always be designed so that the data can be digitally recorded and the experiments can be digitally controlled. The source can be controlled by a computer using an *actuator* or a series of actuators or directly by a current or voltage (as in electrical measurements). The environment is also controlled by a computer using a digital feed back loop. The value of the parameter to be controlled is read by a sensor-digitization system. This value is taken to a computer or controller that checks the measured value with a set value and applies a corrective signal that interacts with the parameter to be controlled by actuators. This is the most basic "architecture" of an experiment as it is controlled by a computer. The computer, sensors and actuators, digitization and digital signal processing are at the heart of modern experimentation. This book has a limited scope of discussing a broad class of experiments. The readers who pursue an advanced career in experimental sciences are requested to consult specialized textbooks in these areas to learn proper applications of these tools. There are a number of sites available in web that specifically discuss digital feed back loop [3].

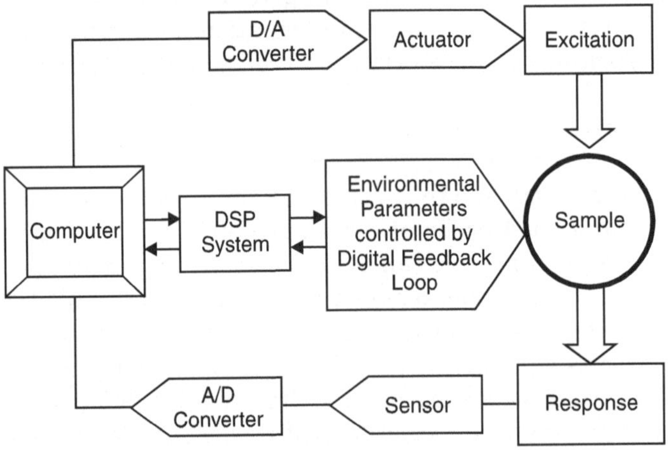

Figure 8.3: Architecture of a modern experiment

In subsequent chapters when we discuss definite experiments we will try to identify the various sensors and actuators used in them, and the control of the experimental parameters with specific examples.

The terms *source* or excitation in Figures 8.1 and 8.3 are used in a general way. For instance when measuring the resistivity the excitation is a constant current. The D/A converter sets a voltage that is converted into a current by a voltage to current converter. The excitation current in this case is set by the main computer as a part of the experimental programme. This arrangement will also contain a digital feed back loop that keeps the current constant. In commercial current sources this is done by an on-board controller and it talks to the main computer by an IEEE interface. But the essential basic concept is the same. Similar is the case of a thermal experiment where the excitation is heat given to a heater. The heater is heated by a current set by the D/A converter and subsequently the current booster. In some experiments these are clearly identifiable and they can be clearly set by the computer. For example in an optical experiment using a monochromator and a broad band source, the intensity of the lamp is controlled by the computer by controlling the current though it by the D/A converter. The monochromator is set at a specific value of wavelength or wave number using an actuator that can be a stepper motor or a linear drive that is controlled by the computer. In experiments where the excitation is a mechanical strain it can be produced by a piezoelectric transducer which is the actuator in this case and which can be controlled by the computer with the D/A card. To summarize, the computer with the help of a D/A card, IEEE 488 interface or with digital I/O outputs can control an actuator that can set the excitation in an experiment.

The response of the experiment is "read" by a *sensor* which gives an output in the electrical domain. This can then be digitized by an A/D card, IEEE interface or digital input (if the sensor gives the data directly in the digital form) and stored in the computer. The stored data can then be processed for obtaining the value of the physical parameter that is being measured. In measurements involving the electrical quantities the sensor is not explicit. The response is a voltage (or current or charge that can be converted to a voltage) that can de directly digitized. In rest of the physical experiments one needs a sensor or a set of sensors that will give an out-put in the electrical domain. In the case of optical experiments the sensor is a photo-diode or a photo-multiplier tube. The former gives a current as the out put which can be converted to a voltage while the latter gives pulses that can be shaped to digital domain for counting. In experiments involving displacement the sensor can be a Linear Variable Differential Transformer (LVDT), a capacitive sensor or an optical system that reflects light from the moving body and make it fall on a position sensitive detector which gives an out put voltage proportional to the displacement of the light spot.

One can then think of the complete experimental arrangement as consisting of three separate but interdependent items. The first component is the "electronics" consisting of the AD or DA cards, I/O cards and such interfaces, computers and amplifiers for the detection or "sensing" end and a current or voltage booster for the actuator end.

An important component of electronics is the software that does the task of data acquisition, and experimental control. There are menu-driven programmes available that can make the software an easy job. A better alternative, however, is to develop a dedicated software in such languages as C^{++} with visual editors.

The second component is the set of sensors and actuators. For example measurement of temperature is done by a thermometer that gives a voltage (in case of a diode thermometer) or a resistance value (in case of a resistance thermometer). The measurement of strain can be done by a strain gauge that gives a resistance value. A piezoelectric positioner or a stepper motor can do positioning of a sample. Success of any experiment and effective utilization of resources depend on understanding the basics of the physical processes of the *transducers* and actuators and also choosing them to optimize performance. Choice of wrong transducers with non-optimal characteristics may be one thing that any successful experimenter would like to avoid. It is important that one reads the characteristics of the transducers and actuators to be used before planning an experiment. As an example suppose we want to do an experiment up to 500 K then the best thermometer to use is a Platinum Resistance Thermometer (PRT). However, if we want to measure down to 1 K or so then a platinum resistance thermometer is no good and we need to use a Cernox or a Ge thermometer. Another example to illustrate this is sample positioning. To position a sample with nanometer accuracy we need a piezoelectric positioner. A stepper motor will not do. On the contrary if we need a long drive for the sample positioning (like a meter or so) with a few tens of micron accuracy we can use a screw drive mechanical system driven by a stepper motor.

The third important component in solid state physics/condensed matter experiment is the material or the sample that is being studied. The sample is produced through various materials growth or processing techniques.

For a good experimenter it is best to have knowledge of all the three components that comprise an experiment. Separating a given experiment into three interdependent yet separate items or tasks makes the experimental design more knowledge based and the job more systematic. This also helps to analyze the performance in a systematic way. Absence of knowledge of one or more of the items converts that part into a "black box". In Table 8.1 we tabulate the above discussion.

Doing the experiment, using digital technology offers several advantages over the analog approach. This includes: speed, accuracy, ease and adaptability. The last point is particularly useful and needs attention. One can build the first two items in Table 8.1 in a modular fashion and around some tested platform so that when a new experiment is being done the changes will be minimal. In many cases such changes can be made with the software. As a thumb rule one can design the experiment in such a way that most of the task is done by the software. This is most desirable as change becomes a change in software.

Table 8.1: The items for the physical experiments

Item/Task	*Implementation methodology*	*Remarks*
Electronics and software	AD *and* DA card, I/O card, computer and software	One can make a general platform and can have software perform all the tasks specific to an experiment.
Signal transduction	Sensors and actuators, Radiation Source	Experiment specific and one can use modular approach.
Sample	Materials preparation	Completely experiment specific; can have certain generic methods.

The concepts used before (namely the excitation and detection along with signal digitization and DSP/computer) are also used for the environmental control through digital feed back loop. The important point is that basic ingredients are the same as the main experiment. The computer or the DSP maintains the environmental parameter with help of a sensor and actuator. The sensor senses the current value of the parameter to be controlled and digital feed back loop gives the corrective signal to the actuator to maintain the value of the parameter at the predetermined value.

4. SENSORS AND ACTUATORS—SOME BASIC ISSUES

In the context of experiment, where the digital control or data acquisition interacts with the analog world is through sensors and actuators. The sensors and actuators that transduce the signal from one physical domain to the other (e.g., pressure to charge) are commonly classified as transducers. A good knowledge of transducers is necessary to obtain the best from an experiment and in particular to optimize the performance of the experimental systems. There are certain characteristics that can be used to quantify the performance of a transducer. This includes Physical dimension, Sensitivity, Linearity, Dynamic response, Range of operation, Hysteresis, Mechanical integrity, Repeatability and Stability. When procuring a transducer, one needs to check these characteristics. In addition one needs to check whether the transducer is compatible with the electronics to be used and whether its sensitivity and response time etc. can give the information one needs to obtain from the experiment. For a good experimental design it is absolutely essential that these estimates are made before an experiment is done. We can take some examples to elaborate this.

We explain the way to make simple estimates using the example of measuring heat capacity. For instance one is measuring the heat capacity c_p of a solid close to the room temperature (~300 K). The sample available is 1 mg. The expected value of c_p for a typical solid near 300 K is in the range of ~500 μJ/K. Suppose the desired precision of the measurement is approximately 1%. In this case one needs to measure the c_p to a precision of 5 μJ/K. For measurement of heat capacity one generally mounts the sample on a substrate (the calorimeter) which carries the heater and the thermometer. Even a small calorimeter made from a sapphire piece of size 5 mm × 5 mm × 0.1 mm (with a thin film heater and thermometer) will have a typical heat capacity $c_s \approx 8$ mJ/K near room temperature. This is much larger than the sample heat capacity. So the resolution in the heat capacity measurement is $x \approx (5\ \mu J/K)/c_s = 6.25 \times 10^{-4}$. The temperature rise after applying a heat pulse, let us say, is $\Delta T \approx 10$ K. (At $T = 300$ K this is an acceptable number). The energy content of the pulse can be measured with a good precision by knowing the power and the time of the pulse. So the experimental resolution will be limited by the precision in the measurement of the temperature rise. The resolution of the temperature measurement needed is $\delta T = x.\Delta T = 6.25 \times 10^{-3}$ K. Suppose one is using a thin film Pt resistance thermometer which has a temperature coefficient of resistivity at 300 K $(1/\rho)\ (d\rho/dT) \approx 4 \times 10^{-3}$K. This is the sensitivity of the temperature sensor. Thus we need to measure a fractional change in resistance of the thermometer $(\Delta R/R) = 25 \times 10^{-7}$. The thin film thermometer can have a resistance R that can be measured and we assume that $R \approx 1$ Kohm. This will need resistance measurement resolution $\Delta R \approx 2.5$ milliohm. Current through the thermometer be I. The Joule heating I^2R due to the measuring current should be much smaller than the heat needed to raise the temperature by δT, which is the order of temperature uncertainty that our resolution asks for. The heat δH needed to cause a temperature rise of the calorimeter by $\delta T \approx 6$ mK is 50 µJ. So the choice of I should be such that $I^2R \approx 50$ µJ. In that case $I \approx 0.2$ mA. With such ac current measuring a resistance change of ΔR will imply measurement of a voltage with resolution $\delta v = I\delta R \approx 500$ nV. When one measures such a voltage even issues like the thermal noise of the 1 Kohm resistor becomes important. It is thus clear that a complete knowledge of the experimental system is of utmost importance when one is doing a precision measurement.

One question arises as to how one can make the design when the actual value of the physical quantity is not known! The way to start is to make a good intelligent guess of the order of magnitude. Do some pilot experiments by using the guess. Then when the rough estimate is obtained, optimize the measurement system parameters and redo the measurement to get the best value.

In general the transducers used are classified into two broad categories—*passive* and *active*. In case of passive transducers one needs an external power source. For example, a resistance thermometer would need a current source for its resistance to be measured. Active transducers do not need any external power source and can directly give a signal in the electrical domain. In Tables 8.2 and 8.3 we give a list of commonly used physical properties that are utilized for making transducers.

Table 8.2: Examples of physical properties that give passive transducers

Physical properties
Resistance and magneto-resistance, Piezo-resistance and photo-resistance
Hall effect
Self and mutual induction
p-n junction
Capacitance

Table 8.3: Examples of physical properties that give active transducers

Physical properties
Piezo-electricity
Thermo-electricity
Pyro-electricity
Photo voltaic

A faithful and reproducible conversion of the electrical quantity to the physical parameter or a faithful conversion of the physical parameter to electrical signal are the most basic issues that one confronts when using an actuator or a sensor. The quality of the actuator or sensor depends on how good or faithfully it does this basic function of conversion of data domain. An important question is what ensures this. This is ensured by the process of *calibration*. Calibration data are often supplied by manufacturers. (When buying a calibrated sensor insist on the traceability of the calibration certificate). These calibrations, however, are not valid for all time. For instance a diode thermometer popularly used by most low temperature experimentalists does not maintain its calibration beyond a couple of years depending on the usage. If this happens then 100 K measured by experimenter *A* will be different from 100 K measured by experimenter *B*. This problem can be avoided by recalibration. Thus the sensors and actuators used need periodic calibration with an in-house facility or by a calibration laboratory that is accredited. In India National Accreditation Laboratory Board (NABL), an organization of the Department of Science and Technology (DST), does this accreditation. It is good to send sensors, actuators and the measuring equipments for periodic calibration to the accredited laboratories to ensure that the experiment is properly bench-marked. An experimentalist should be aware of this issue and should note that experiments cannot be bench-marked without proper calibration. Table 8.4 given below shows a list of different primary standards used in physical measurements. The other physical quantities are deduced from these physical quantities using relevant physical theories.

Table 8.4: The primary standards for the physical parameters

Parameter	*Primary standard used for calibration*
Voltage	10 V Josephson Junction Arrays
Length	Krypton wavelength and Laser Interferometer
Resistance	Quantum Hall Effect Standard
Time and Frequency	Atomic Clock (*Cs*) and Laser cooled atoms
Temperature	Calibrated to International Temperature scale. He^3 melting curve, Vapor pressure of He^4, Gas thermometers, Platinum resistance thermometer, Fixed points like Melting temperature of pure metals, Triple point cells and Radiation thermometers based on block body curves.
Pressure and vacuum	Air dead-weight tester and precision manometers
Mass	Standard Kg

5. NOISE FUNDAMENTALS

In a physical experiment we measure an electrical quantity v_o (say a voltage) corresponding to the physical parameter X_0. Let us call v_o the true value. The job of the experiment is to ensure that the measured average value $<v>$ is as close to v_o

as possible so that $<v>$ faithfully represents X_0. We quantify the departure as an error in measurement. The absolute accuracy (referred to as accuracy) is the degree of agreement between the true value and the measured value. The relative accuracy (referred to as precision) is the degree of repeatability or resolution.

In an actual experiment the error arises from a number of sources. Let the signal measured by the experiment be represented by the time series $v(t)$. This is made up of two components:

$$v(t) = s(t) + n(t) \tag{8.5.1}$$

where $s(t)$ is the signal and $n(t)$ is the noise. The error in measurement is contributed by $n(t)$. There are three main contributions to $n(t)$. They are: (*i*) instrumental contributions, (*ii*) contributions inherent in the system like those from thermal noise and flicker (1/*f*) noise and (*iii*) contributions from external sources which we collectively call *noise*. In this chapter we discuss the likely sources of *noise* and how to do noise reduction by hardware methods. As can be seen from eqn. 8.5.1, the measured signal can be strengthened by reducing $n(t)$ with respect to $s(t)$, a process known as Signal to Noise enhancement. Often digital techniques are extensively used for *S/N* enhancement.

One of the important tasks in experimental physics is to enhance the signal over the noise. This strength of the signal (or the cleanliness of the experiment) is quantified through what is called *Signal-to-Noise (S/N) ratio*. The attempt of any experiment is to maximize the *S/N* ratio. This requires a number of steps needing both software and hardware. Though a good deal of the *S/N* ratio enhancement can be done by using various digital techniques such as digital filtering, correlation techniques, Lock-in techniques etc., it is always best to make the hardware clean enough so that the maximum benefit can be obtained from the *S/N* enhancement technique. Some conventional wisdom is widely used to clean the measurements and enhance the signal. In this chapter we will discuss some of these common techniques and will refer to texts that specifically deal with them.

As discussed above there are two aspects to *S/N* enhancement. The first is to reduce the noise by hardware methods and other is the use of digital techniques to enhance *S/N* ratio. In this section we discuss the hardware aspects and in the next chapter we will discuss the digital techniques. Reduction of noise in an experiment is not a "black magic". It can be based on knowledge and the knowledge can be made effective.

How do we know, from the time series $v(t)$, what is signal and what is noise? This is definitely a crucial question and if there is no intelligent way to distinguish the two one can retain the noise mistaking it as the signal and thus increase the error. In the other extreme, one may filter out the signal mistaking it as the noise thereby distorting the experimental outcome. The best way to distinguish these two is to look at the *power spectrum* of the time series in the frequency domain. Generally, the signal and noise will have different power spectra. This helps us to design digital filters to suppress the noise. We discuss this in the next chapter.

Here we briefly state how the power spectra can distinguish the signal from the noise. Power spectrum of the time series $v(t)$ is written as $S_v(f)$ and for any physical quantity X it is defined as:

$$S_X(f) = \underset{T\to\infty}{Lt}\left(\frac{1}{2T}\right)\left|\int_{-T}^{+T} dtX(t)e^{-i2\pi ft}\right|^2 \tag{8.5.2}$$

The two parts of $v(t)$ namely $s(t)$ and $n(t)$ are generally statistically independent and hence they will make independent contributions to the power spectrum. If these two contributions make distinctly different contributions in $S_v(f)$ one can separate these two parts easily. We illustrate this in the Figure 8.4 below.

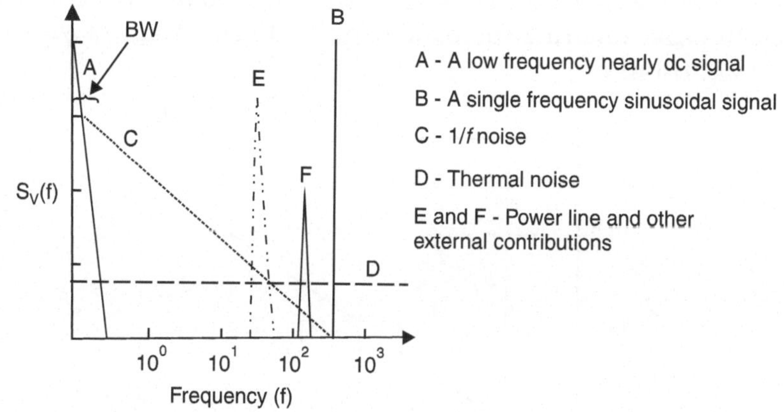

Figure 8.4: Contributions by different components in the power spectra

If we have a very low frequency or dc signal, then the $1/f$ noise contributions, mainly from the amplifier and other electronics, will make the overwhelming contributions and can submerge the signal part unless the band width (BW) of the detector is limited to the region marked by the arrow in the Figure 8.4. This is essentially low pass filtering. The line power and such extra sources make discrete contributions marked E and F in the figure. If the signal is occurring in this region of frequency it will be severely corrupted. A knowledge of the signal frequency is thus very important. Some times by mixing the signal with a low frequency one can move to a region of frequency (like the signal marked B) which is away from the interference range and the $1/f$ noise also makes a small contribution.

The fundamental limit to measurement is the Johnson noise (also known as Nyquist noise or thermal noise). This has a frequency independent white power spectrum as shown by the line D in Figure 8.4 [4]. The spectral power due to *Johnson noise* for a resistor of value R ohm kept at a temperature of T Kelvin is:

$$S_V(f) = 4k_BTR \qquad V^2.Hz^{-1} \tag{8.5.3}$$

$V^2.Hz^{-1}$ is volt^2/Hz.

The contribution to the mean square voltage noise due to this spectral power within a *bandwidth* B is given as:

$$\langle V^2 \rangle = \int_{f_{\min}}^{f_{\max}} df\ S_V(f) = 4k_B TR\ B \qquad (volt^2) \tag{8.5.4}$$

where $B = f_{\max} - f_{\min}$ is the bandwidth. The spectral power due to the thermal noise can be rather large in a large resistance even with moderate bandwidth. In Figure 8.5 we show the rms noise voltage due to thermal noise in source resistance of 1 Kohm kept at a temperature of 300 K. For each decade in resistance increase, the rms noise goes up by a factor of $\sqrt{10}$. We can see that even at moderate bandwidth of 100 KHz the rms noise for a source resistance of 1 Kohm can be in the range of 1.3 μV peak to peak. Limiting the bandwidth to 10 Hz brings down the thermal noise to below 0.013 μV.

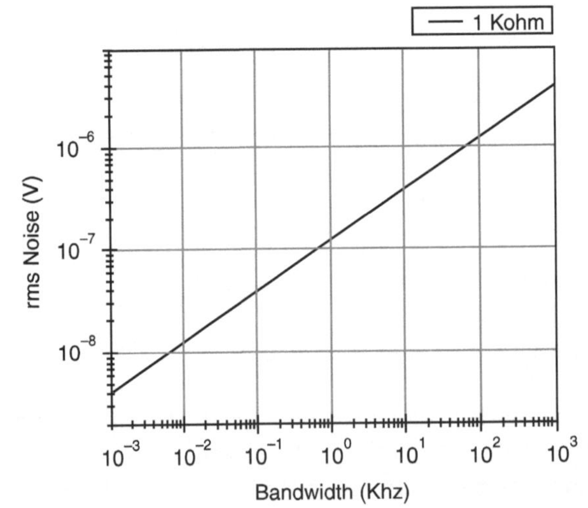

Figure 8.5: Bandwidth limited thermal noise from a 1 Kohm resistor

For 1 G-ohm and only 1 KHz bandwidth this can be as high as 0.13 mV peak to peak. One can immediately see the effect of limiting the bandwidth on thermal noise. But B cannot be reduced arbitrarily as it affects the response of the circuit. The bandwidth B is often determined by the highest frequency limit of the detection circuit. Typically as a thumb rule if the 10%-90% rise time of the instrument is t_R (sec) than the bandwidth (in Hz) $B \approx \frac{0.55}{t_R}$. We can immediately see that to have at least a $t_R \approx 1$ m sec, $B \approx 0.55$ KHz. This is a finite bandwidth. If the detection is by a *AD* card then the bandwidth is $B \approx \frac{0.314}{t_{INT}}$, where t_{INT} is the time of integration of the *AD* card. For a high resistance circuit often the *RC* time constant determines the bandwidth by the input capacitance C_{IN} such that $B \approx \frac{0.25}{\tau}$, where $\tau \approx R.C_{In}$. Generally to reduce the thermal noise we reduce the bandwidth by low pass filtering. But there is a limit to that because of a possible severe reduction of the response time of the experiment.

5.1. Noise Pick-up by Inductive and Capacitive Coupling

The extraneous noise that interferes with the measurement generally arises from a source (for example a pump running in the room, a FM radio playing or a flickering tube light) and there is a coupling path that couples the noise source to the experiment. As an example inductive coupling or ground loop provides a good coupling path for electrical noise. A strong coupling of the mechanical noise source with the experiment is often provided by a concrete floor or a structure that can resonate with the frequency of the mechanical source. Noise reduction is best done by eliminating (if this is not possible, at least by significantly reducing) both or at least one of the two—namely the source as well as the coupling path. Most of the time it is difficult to eliminate the source of noise. So it is best to reduce or eliminate the coupling path. In the context of electromagnetic interference this is done through grounding and shielding. But electromagnetic interference is not the only source of noise. In some sensitive experiments, particularly those involving precision positioning and spatial resolution approaching nm or better (as in modern scanning microscopes), even a building vibration can be the source of the extra noise. Damping mechanisms can eliminate this. In this case the damping provides a reduction of the coupling path to the source of the noise. Some times thermal sources can be cause of noise. For instance the experiment is being carried out in a temperature-controlled environment. If there is a sufficient fast change in temperature to which the control loop cannot adjust, this will show up as a noise in the measurements. In systems that have a feed back loop controlling the physical parameters a noise source is the feed back loop itself! Depending on the feed back loop time constant and the time scale of change one may get a contribution coming from the environment of the experiment. This happens because if the loop is not properly tuned then the oscillations and overshoot in the feed back loop will introduce instability in the value of the experimental parameter being controlled and this will couple to the experiment directly. This can be checked by checking the feed back loop as the experiment is going on, and then can be suitably corrected from the recorded experimental value. In some experiments even these can be very serious. For instance in some low temperature experiments which are very sensitive to the magnetic field one can get noise coming from even the magnetic field of the current that controls the temperature! A very interesting source of noise is often encountered in ac measurement of resistance in a magnetic field. Due to the alternating current in the leads to the sample, the leads will experience a Lorentz force due to the magnetic field and thus will oscillate at the frequency of the alternating measuring current. The finite amplitude of oscillation will encompass a finite area leading to an induced voltage that will appear as a spurious signal in the quadrature. In the following part of the chapter we discuss the electrical noise which is the most common component of noise that an experimentalist encounters. For a complete discussion we give reference for a more comprehensive treatment [5].

Methods of coupling to the noise source are mostly: (*i*) Electromagnetic coupling either by inductive or capacitive routes, (*ii*) Common impedance coupling, (*iii*) Conductive coupling. Additional sources of noise come from (*i*) lack of mechanical stability, (*ii*) choice of wrong materials in circuit design and their usage and (*iii*) environmental factors.

Electromagnetic coupling is all-pervasive. Due to the proportionality of the induced voltage on the frequency it is very severe in high frequency measurements. The coupling of noise to experiment is generally reduced by grounding and shielding and using certain basic techniques like avoiding common impedance to a noise source. For details of noise coupling to the experiment and its reduction the reader should refer to specialized texts. In this book we provide a brief resume that will give some elementary knowledge to an experimenter. Noise reduction is neither a black magic nor is it a case of trial and error. This is based on sound principles and a good experimenter should acquire this knowledge and apply it to the extent possible.

Electromagnetic fields radiated by a source, which is a part of the experiment, or an external source, provide a means of noise. Many-a-time this can overwhelm an experiment. A flickering tube light can couple enough noise power (~few tens of μW) in a low temperature experiment to cause warming. A radio-set working near-by can cause enough interference to an experiment where one is measuring signals of tens of nV. Electromagnetic radiation is the most common source of noise in most experiments because any fluctuating current or voltage (any moving charge to be precise) can cause this radiation. This mode of coupling is best avoided by shielding and grounding that reduces the coupling strength. The capacitive and inductive couplings are the two paths that introduce noise in the experimental system.

In Figure 8.6 we show with a cartoon the basic concepts of capacitive coupling. The unshielded portions of the two circuits contribute to the capacitance C_c. The pick-up voltage is given by:

$$\left|V_{pickup}\right| = \omega R C_c V_a \tag{8.5.5}$$

The coupling capacitance for a pair of wires of diameter d and unshielded length L kept at distance D is given by $C_c = \left[\pi\varepsilon L/ln(2D/d)\right]$ Farad. The pick-up voltage increases as the frequency increases and it reduces with decrease of the detector resistance R. Thus a high resistance input often becomes a source of large noise pick-up unless properly shielded against capacitive pick-up.

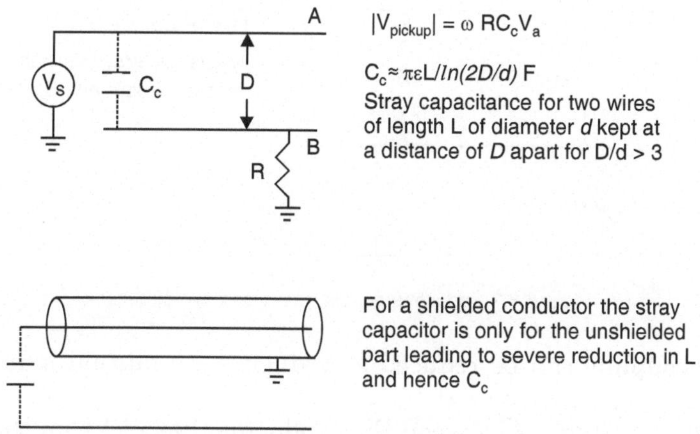

Figure 8.6: Capacitive coupling and effect of shielding

The following examples give the extent of voltage pick-up. Let us consider two wires of diameter 1 mm and length 1 m placed parallel at distance of 10 cm. One of the wires is carrying an ac signal of peak-to-peak voltage of 1 V at a frequency of 100 KHz. The other wire, which is in the detector side, is terminated with a resistance of 1 Mohm. This is the detector effective resistance. In that case the stray capacitance is around 6pf and the noise pick up voltage is around 3.8 V peak to peak. This is larger than the voltage in the source! Even for a detector resistance of 1 kOhm the pick up can be in the mV range. Given the high input impedance needed in a number of experiments even a pF order of C_c can lead to substantial pick-up once the frequency crosses few tens of KHz.

Reducing C_c reduces capacitive coupling. Keeping a grounded shield between the two wires as is done in a coaxially shielded wire can best do this. In this case the stray capacitance is only due to the exposed wire as shown in the figure and this can reduce the pick-up by orders depending on the shielded length of the wire. Another way to reduce the stray capacitance is to make the two wires perpendicular. This however is not always possible. Increasing the separation does not reduce C_c much because the log dependence on the distance.

More severe source of noise pick-up is through the inductive coupling. In fact in many cases there are number of unattended sources of inductive coupling that can cause uncontrolled noise pick-up. Most important and often nontrivial source of noise pick-up is the return path of the current through the device to the ground and wrong use of the shield. In Figure 8.7 we give a cartoon of the inductive pick-up.

The induced voltage in the circuit is given by the relation:

$$\left|V_{pickup}\right| = \omega M_{ab} I_a \tag{8.5.6}$$

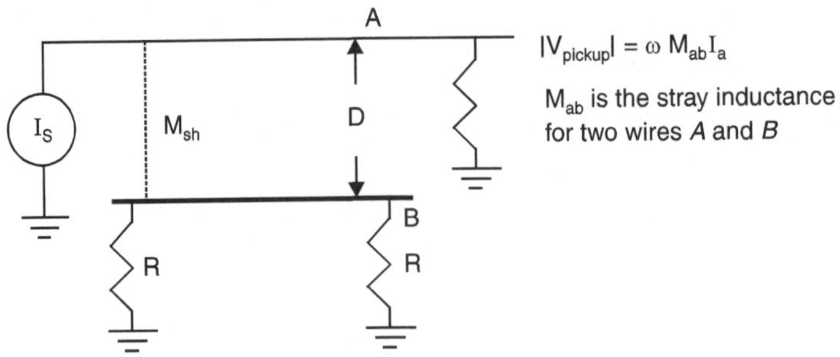

Figure 8.7: Inductive coupling between two circuits

Inductive coupling can be reduced by reducing the mutual inductance M_{ab} of the source and the pick-up wire (which we call the receiver). In case of capacitive pick-up the noise voltage is between the conductor that picks up the voltage and the ground. In case of the inductive pick-up this noise voltage appears in series with the pick up conductor.

Another way we can evaluate the pick-up voltage is the average magnetic field <*B*> created by the interfering source (called source here) at the receiver over its average area *A*. The average pick-up voltage $V_{pickup} \approx -\frac{d(\langle B \rangle A \cos\theta)}{dt}$, where θ is the angle between <*B*> and *A*. We can clearly see that the pick-up can be reduced by reducing either or both <*B*> and *A*. <*B*> can be safely reduced by shielding the current carrying wire and making the return current flow through the shield, or by a twisted pair both of which ensure that the magnetic field produced by the current through the forward and return path move almost in the opposite direction there by canceling the magnetic field and thus severely reducing <*B*>. Similar considerations apply for the receiver side where the loop area *A*, for the pick-up, can be severely reduced by having a shield (which is properly grounded) around the central conductor. This makes the onward and return path close together thus reducing the area A. Similar thing happens for the twisted pair. Another important way the pick-up voltage can be reduced is through increase of the angle θ to 90° thus minimizing the pick-up. To summarize, inductive coupling crucially depends on the return path of the current and how close it is to the onward current. If they are physically close and flow in opposite directions, this will reduce <*B*>. And on the receiver side this will reduce the loop area. In addition to this consideration, there is a crucial parameter that depends on frequency ω. In the reduction of magnetic (or inductive) coupling, the current through the shield is an important factor, which we mark as I_s. The induced voltage V_s on the shield will drive the shield current that will be determined by the impedance Z_s of the shield. At frequency ω, $Z_s = R_s + j\omega L_s$, where R_s is the shield resistance and L_s is its inductance. The current through the shield I_s for current *I* in the central conductor is given by:

$$I_s = \left(\frac{j\omega}{j\omega + \omega_s}\right) I \tag{8.5.7}$$

The shield has a frequency cutoff at $\omega = \omega_s \equiv \frac{R_s}{L_s}$. For frequency $\omega > \omega_s$, $I_s \approx I$ and the return path of the current will be through the shield thus providing effective shielding. At lower frequency, $I_s \propto \omega$ and as $\omega \to 0$, $I_s \to 0$. In this case most of the return path will be through the ground path thus reducing the effectiveness of the shielding.

6. ELECTRIC AND MAGNETIC SHIELDING

One of the effective ways one can reduce pick up is by shielding. The Physics of shielding is an application of Maxwell's equations and simple ideas can be used to evaluate it. Often to cut down on external noise one carries out experiment in a shielded room. The strength of shielding is measured by the ratio

$$S(dB) \equiv 20 \log\left(\frac{E_0}{E_1}\right) = 20 \log\left(\frac{H_0}{H_1}\right)$$

called the shielding factor where E_0 and H_0 are the electric and magnetic fields outside the enclosure and E_1 and H_1 are the corresponding fields inside. The shielding factor can be calculated when designing an enclosure for experiment. The total factor can be decomposed into a reflection component (R) and an absorption component (A) that arises due to skin depth so that $S = R + A$. The reflection loss can be evaluated by a simple method utilizing impedance for propagation of electromagnetic (em) waves. Physically when an em wave (E_i, H_i) falls on a metal from a vacuum the transmitted intensity of the electric and magnetic fields are

$$E_t = \frac{2Z_m E_i}{Z_v + Z_m}, H_t = \frac{2Z_v H_i}{Z_v + Z_m} \tag{8.6.1}$$

where the impedance of vacuum is Z_v and the impedance of the metal is Z_m. In far field approximation, the impedance $Z_v \approx 137$ ohm while for metal $Z_m = \sqrt{\frac{\omega\mu}{\sigma}}$. For a metal like Cu, $Z_m = 3.68 \times 10^{-7} \sqrt{f}$ ohm. Even up to a frequency of 100 MHz, $Z_m = 3.68 \times 10^{-3}$ ohm and thus $Z_m << Z_v$. From equation (8.6.1) we can see that $E_t \cong 2\frac{Z_m}{Z_v} E_i$ while $H_t \cong 2H_i$. Thus the first surface provides the electric shielding by reflection loss. Thus even a thin metal plate can provide effective electric field shielding. In case of a thick shield (thickness >> skin depth), however, both the fields are equally attenuated and we have:

$$E_t, H_t = \frac{4Z_m Z_v (E_i, H_i)}{(Z_v + Z_m)^2} \tag{8.6.2}$$

Expressing the impedance for a metal with relative conductance (expressed with respect to Cu) $\sigma_r \equiv \frac{\sigma}{\sigma_{Cu}}$ and relative permeability as $\mu_r \equiv \frac{\mu}{\mu_{Cu}}$ as $Z_m = 3.68 \times 10^{-7}\sqrt{f}\sqrt{\frac{\mu_r}{\sigma_r}}$ we find the reflection loss in dB as:

$$R(\text{dB}) \equiv 20\log\left(\frac{94.25}{3.68\times 10^{-7}\sqrt{\frac{\mu_r f}{\sigma_r}}}\right) \cong 168 - 10\log\left(\frac{\mu_r f}{\sigma_r}\right) \tag{8.6.3}$$

Equation (8.6.3) gives an interesting result that for reflection loss the maximum loss is 168 dB at $f = 0$ independent of the metal and the loss decreases as f increases. Thus shielding at low frequency is most effective due to reflection loss from a metal.

The absorption loss is totally decided by the skin depth $\delta \equiv \sqrt{\frac{2}{\omega\mu\sigma}}$. The absorption loss in dB for a metal sheet of thickness t is given as:

$$A(\text{dB}) \equiv 20\log_e\left(\frac{t}{\delta}\right) \cong 8.69\left(\frac{t}{\delta}\right) = 131t\sqrt{f\mu_r\sigma_r} \tag{8.6.4}$$

where the thickness is in meters. The loss increases as the frequency increases and it is most effective in magnetic materials like steel and even better in materials like mu-metal that can have very high relative permeability (higher by few orders) although sacrificing by a factor in the conductivity. Briefly, the noise shielded enclosure can be built by a bi-layer combination where the outer layer provides the reflection loss by a good conductor like Cu and then an inner layer of steel or mu metal with a thickness which is greater than the skin depth in the frequency range of operation.

REFERENCES

1. Chaikin P.M. and Lubensky T.C., "Principles of Condensed Matter Physics" (Cambridge University Press) (2000).
2. Lee P.A. and Ramakrishnan T.V., Rev. Mod. Phys. **57**, 287 (1985).
3. Websites; www.engin.umich.edu/group/ctm/digital and www.ni.com/pdf/academic.
4. Reif F., "Fundamentals of Statistical and Thermal Physics" (McGraw Hill) (1965).
5. Henry W. Ott, "Noise Reduction Techniques in Electronic systems" (Wiley-Interscience) (1998) and Ralph Morrsion "Grounding and Shielding Techniques" (Wiley-Interscience) (1998).

9 CHAPTER

Basic Digital Technique Concepts

1. INTRODUCTION

Almost all present day experiments use digital techniques. In this chapter we will introduce the reader to some of the basic ingredients needed to understand how digital techniques are applied particularly in two areas namely signal to noise ratio enhancement and digital control of feed back loop. In this chapter our attempt is to provide a rudimentary yet working knowledge of this modern area. The first important task of any digital technique is to convert the signal which is an analog signal to digital signal (a set of discrete numbers) that are stored in the computer for further processing which is collectively known as Digital Signal Processing (DSP) [1].

2. DIGITIZATION OF ANALOG SIGNAL

Digitization involves two steps: (1) The analog signal is *sampled* at finite interval of times (see Figure 9.1(a)) and (2) Each sample is quantized into a finite number of digits (see Figure 9.1(b)). The operation of digitization is done by an *Analog-Digital converter* card (ADC) that sits on a specific bus of a slot of the computer (like the PCI bus of a Pentium). Commercially available AD cards are sold by specifications on sampling rate and bits for digitization. The 200 KHz/16 bit card is a rather common one [2]. For lower bits like 12 bit, the speeds are much higher like one MHz or more. The cartoon below explains the two steps.

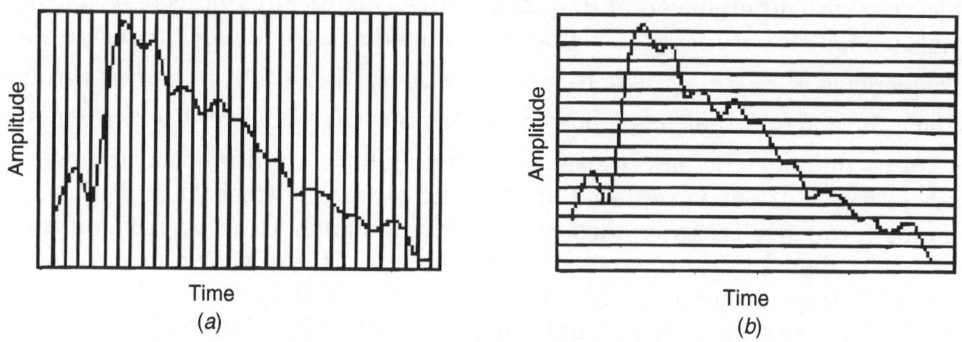

Figure 9.1: (a) Digitization in time domain and (b) Quantization of an analog signal

The conversion of analog signal $V(t)$ to a set of digitized numbers sampled at regular intervals (T) is represented by $\{V_n\} \equiv \{V_0, V_1, ... V_n\}$ where the data have been collected over a time NT. Both these quantities T and NT are determined by the frequency characteristics of the signal and will thus determine whether the digitized signal contains the necessary information that is there in the data. Sampling time T (and hence the rate $1/T$) will determine the bandwidth and the maximum frequency component of the data. *Band-width* B of the signal and the sampling frequency or rate $f_s \equiv 1/T$ are related by the *Nyquist theorem* [3]. The f_{min}, the minimum frequency that the sampled signal can access will be determined by the total length of the sampled data and $f_{min} \approx 1/NT$. Longer we sample (larger N) smaller is the f_{min} that we can obtain. The set of numbers $\{V_n\}$, to be a true replica of the analog signal, should contain as much relevant information on the original signal as practically possible. Very often improper digitization can lead to wrong results and undesirable noise in the data.

The sampling rate $1/T$ is dictated by the Nyquist theorem which states that if a signal is band-limited, i.e., there is a lower limit f_{min} and an upper limit f_{max} of frequency components in the signal, then the sampling rate $1/T$ should be such that $1/T \geq 2B$. When the signal is sampled at a frequency $f_s < 2B$ the signal is *under sampled* and there is loss of information and this leads to aliasing at low frequency, which we will discuss briefly. When $f_s > 2B$, one can use the interpolation function $g(t)$ to obtain back the analog signal $V(t)$ from the $2N$ sampled points $\{V_n\}$, where $V_n \equiv V(nT)$ so that:

$$V(t) = \sum_{n=-N}^{n=+N} V_n g(t - nT) \tag{9.2.1}$$

where the interpolation function $g(t)$ is given as:

$$g(t) = \frac{\sin(2\pi Bt)}{2\pi Bt} \tag{9.2.2}$$

For a signal characterized by the bandwidth, an important consequence follows from *Nyquist theorem*. Two signals with the same spectral power and bandwidth, but located with different center frequency, can be identically sampled. A continuous signal component with frequency f, when sampled at a frequency f_s, creates *aliases* at frequencies $F_m = f \pm mf_s$ where m is an integer, Figure 9.2 shows this through a cartoon, for a signal that has a power spectrum centered around zero frequency.

After sampling, the power spectra of signals have similar spectral power and bandwidth B. But one is centered at zero and others centered at multiples of the scanning frequency creating alias of the original signal. Thus we have an infinite numbers of aliases that are created by the process. This is known as *aliasing*. For a properly sampled signal ($f_s > 2B$) the aliases are well separated and one can get a proper reconstruction of the signal using an appropriate interpolation function, like that given in equation 9.2.2. Obviously below the critical sampling rate $2B$, the aliases overlap and reconstruction of the signal leads to inaccuracy. This is illustrated in cartoon shown in Figure 9.2. Failure to satisfy the Nyquist requirement will result in *aliasing* of higher-frequency components, meaning that these components will appear to have frequencies lower than their true values. One way of avoiding the problem of aliasing is to apply a low-pass filter to the signal, prior to the sampling stage, to remove any frequency components above the "folding" or Nyquist frequency (half the sampling frequency). Such *anti-aliasing filters* are commonly built into the analog interface chips and codecs that convert analog input signals into digital form for processing by a digital signal processor (DSP). In many cases, these anti-aliasing filters are implemented using conventional analog circuitry. An alternative method is to use a digital anti-aliasing filter. This avoids the noise and drift problems inherent in analog filter circuits, and is the natural approach when the signal is going to be processed digitally anyway. MATLAB® allows design of anti-aliasing filters.

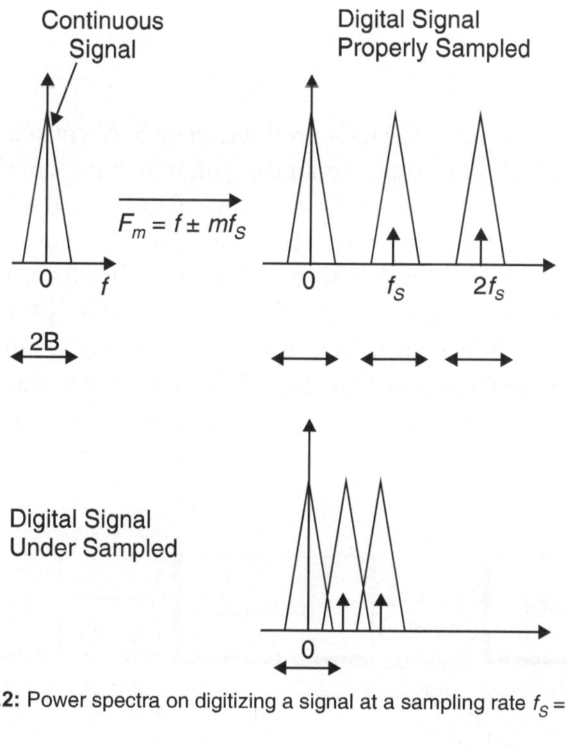

Figure 9.2: Power spectra on digitizing a signal at a sampling rate $f_s = 1/T > 2B$

Quantization of the analog signal is determined by the number of bytes the AD card can accommodate. The process of converting an analog signal to discrete numbers gives rise to what is known as *Quantization error* or Quantization noise. Suppose we are digitizing with a 16-bit card. Suppose the maximum signal the card can take is 0-1 V. In that case the continuous analog voltage is mapped into one of the 2^{15} = 32768 values with minimum separation of 1/32768 ≈ 30 μV. Quantization thus introduces a rounding off error. The Quantization noise is defined as the difference between the actual value of the analog signal, for the particular sampling time, and the nearest quantization interval value. At most, this error can be as much as half of the interval. The quality of the quantization is characterized by the *Signal to Quantization Noise Ratio* (SQNR). For a N-bit Quantization per sample the maximum signal is mapped to $2^{N-1} - 1 \approx 2^{N-1}$ points and the most negative signal to -2^{N-1}. We can define the SQNR as

$$\text{SQNR(dB)} \equiv 20\log_{10}\frac{V_{signal}}{V_{Quantization}} \cong 20\log_{10}\frac{2^{N-1}}{(1/2)} = 6.02\,\text{N} \qquad (9.2.3)$$

One can clearly see that SQNR increases with N and per extra bit it goes up by nearly 6 dB. In short, the digitization rate and the number of bits (or SQNR) define the basic characteristics of a digitization process. In general as the speed goes up the number of bits goes down.

3. DIGITAL FILTER

One of the most common tools for noise reduction or *S/N* ratio is filtering using low pass, band pass and high pass filter depending on which part of the power spectrum of the signal we want to retain. Filtering has also been a viable tool in analog signal processing. DSP allows exotic filter design with software that is almost impossible to implement by analog filters. Filtering is often, though not always, done to suppress noise. It depends on the signal's frequency spectrum being different from that of the noise. Filtering is a process of selecting, or suppressing, certain frequency components of a signal. The digital filter allows to pass certain frequency components of the signal; in addition it can attenuate, or suppress, each frequency component by a desired amount. Thus a digital filter can shape the frequency spectrum of the signal.

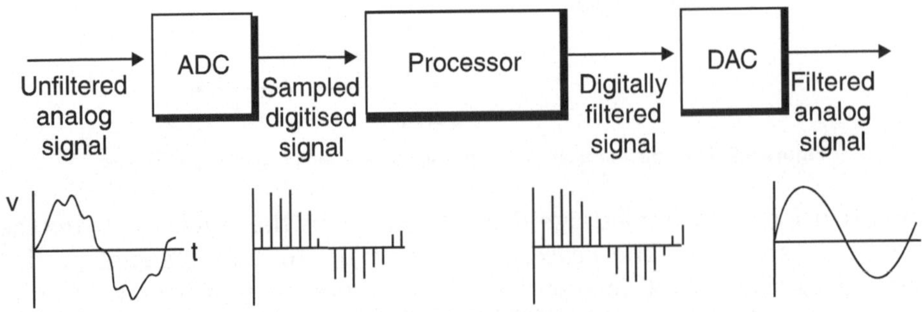

Figure 9.3: Basic block diagram of implementation of a digital low pass filter

The cartoon (Figure 9.3) describes the basic process of digitizing a noisy signal by AD converter (ADC), processing the signal through a digital low pass filter and then reconstructing this by using a DA converter (DAC).

Suppose we want to low pass filter a signal $V(t)$. There are two ways we can filter the signal. We can evaluate the Fourier transform of the signal which is $\tilde{V}(f)$ and then we multiply the Fourier transform of the signal by the filter function $H(f)$. We get the filtered signal $Y(f)$ in the frequency domain. We can then take the inverse Fourier transform to get the filtered signal. This process clearly is computation intensive because it involves two Fourier transforms. This is given in the form of equation (9.3.1) below.

$$V_{filtered}(t) = \int_{-\infty}^{+\infty} dfe^{-j2\pi ft}\left\{\tilde{V}(f) \times H(f)\right\} \tag{9.3.1}$$

Alternatively we can use the convolution theorem, which states that Fourier transform of the convolution of two functions is the product of their Fourier transforms:

$$(\tilde{f} \otimes \tilde{g})\,(x) = FT \int_{-\infty}^{+\infty} f(t)g(x-t)dt \tag{9.3.2}$$

where the left hand side is the product of the Fourier transforms and right hand side is the Fourier transform of the convolution. Using the convolution theorem, the filtered function in the time domain can be obtained by convoluting the time domain signal with the Fourier transform of the filter function $\tilde{H}(t)$. This is given in equation (9.3.3). This is generally the route adopted in designing a digital filter:

$$V_{filtered}(t) = \int_{-\infty}^{+\infty} dfe^{-j2\pi ft}\left\{\tilde{V}(f) \times H(f)\right\} = \int_{-\infty}^{+\infty} V(x)\tilde{H}(x-t)dx \tag{9.3.3}$$

Digital filters are also characterized by their response to an impulse. The *impulse response* is an indication of how long the filter takes to settle into a steady state. It is also an indication of the filter's stability—an impulse response that continues oscillating in the long term indicates the filter may be prone to instability. A widely used digital filter is the *Finite Impulse Response* (FIR) filter. It is much easier to approach the problem of calculating filter coefficients if we simplify the filter equation so that we only have to deal with previous inputs (that is, we exclude the possibility of feedback). If such a filter is subjected to an impulse (a signal consisting of one value followed by zeros) then its output must necessarily become zero after the impulse has run through the summation. So the impulse response of such a filter must necessarily be finite in duration. It is for this reason such a filter is called a Finite Impulse Response filter.

The digitized voltage signal is given by the set $\{V_n\}$ and the Fourier transform of the Filter function $\tilde{H}(t)$ is also represented by a series of filter coefficients $\{\tilde{H}_m\}$. In this case the filtered digitized voltage will be represented by the set $\{Y_n\}$ that is given as:

$$Y_n \equiv \sum_{m=-N}^{N} V_{n-m} \tilde{H}_m \tag{9.3.4}$$

2N + 1, the number of terms in the filter function, is its order or width. A continuous filtered voltage can be reconstructed from the set $\{Y_n\}$ by using equation (9.3.1).

This is a finite impulse response because filter is non-recursive. In this case the filter has an order 2N + 1. Often the order is referred to as the *width of the filter function*. The filter's frequency response is obtained from the filter coefficients $\{\tilde{H}_m\}$ and is given in this case as:

$$H(f) = \sum_{m=-N}^{N} \tilde{H}_m e^{-j2\pi m f/N} \tag{9.3.5}$$

A very important parameter in filter design is how many filter coefficients to be retained. To overcome the problem of large number of filter coefficients that are needed to sharpen the response of the filter function, FIR filter coefficients can be calculated using the window method using *window functions* [4].

An example of the window function is the *Hanning window function* defined as:

$$\tilde{H}_m \equiv 0.5 + 0.5\cos(2m\pi/N) \tag{9.3.6}$$

for $-N/2 \leq m \leq N/2$; else $\{\tilde{H}_m = 0\}$. The filter function in the frequency domain is shown in Figure 9.4. N is the length (or width) of the filter.

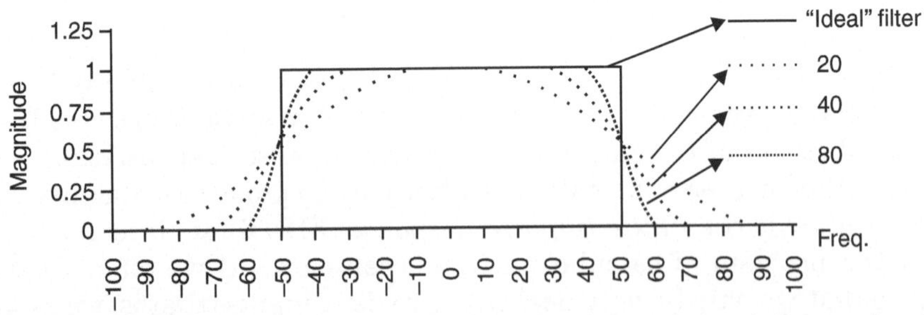

Figure 9.4: Hanning Window (with varying lengths N = 20, 40, 80) in frequency domain in comparison to the ideal filter

The finite number of terms suppresses the sharp response of the ideal filter and as the number of terms increase, this brings the response closer to the ideal filter function.

An effective way to enhance the *S/N* ratio is averaging of the data. This actually is low pass filtering. For instance, for a "three term average filter" where the average is taken of the current and two previous inputs, the filtered function is defined as:

$$Y_n \equiv \frac{V_n + V_{n-1} + V_{n-2}}{3} \tag{9.3.7}$$

The filtered values Y_0 (and Y_N) need values like V_{-1}, V_{-2} and (V_{N+1}, V_{N+2}). These values are taken as zero. In the above example of three term average, the calculation of current out put Y_n depends on two previous inputs such as V_{n-1} and V_{n-2}. The above example although trivial defines a basic filter.

Digital filters can be specified (like analogue filters) by the pass-band, stop-band, stop-band attenuation, pass-band ripple and stop-band ripple as explained in Figure 9.5 below. Pass-band is the band of frequency components that are allowed to pass. Stop-band is the band of frequency components that are suppressed. Pass-band ripple is the maximum amount by which attenuation in the pass-band may deviate from nominal gain. Stop-band attenuation is the minimum amount by which frequency components in the stop-band are attenuated.

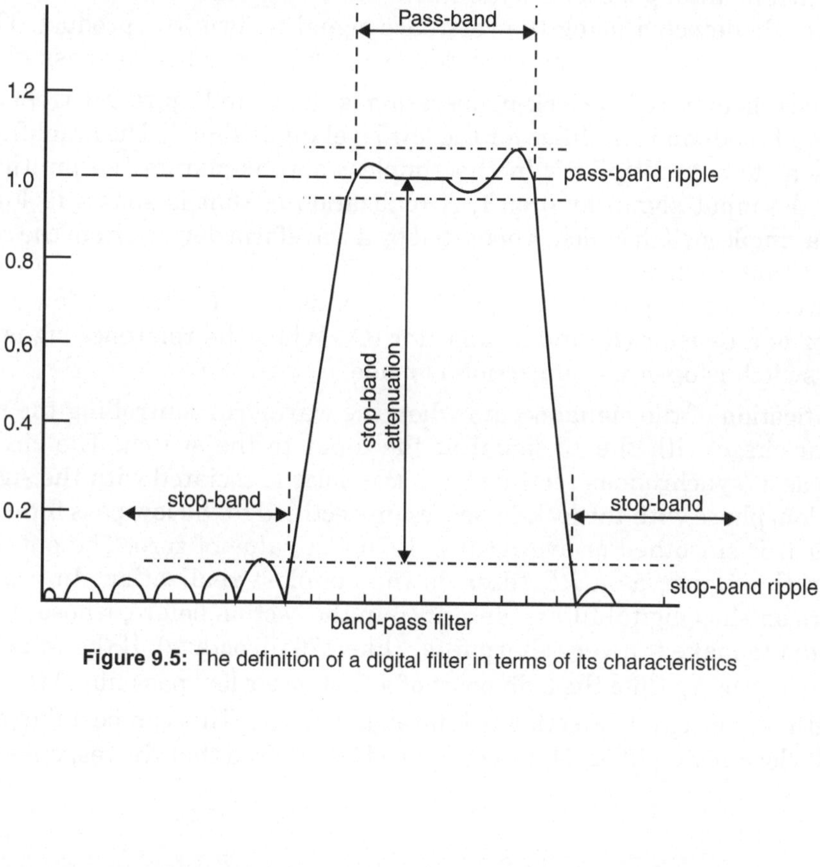

Figure 9.5: The definition of a digital filter in terms of its characteristics

4. PHASE SENSITIVE DETECTION AND LOCK-IN AMPLIFIER

One of the widely used methods for *S/N* enhancement is phase sensitive detection. This is implemented in a much-used equipment called *Lock-in amplifier*. Stated simply, *phase sensitive detection* refers to the demodulation or rectification of an ac signal by a circuit that is controlled by a reference waveform derived from the device that caused the signal to be modulated. The phase sensitive detector effectively responds to signals that are coherent (same frequency and phase) with the reference waveform and rejects all others.

As discussed in section 8.5 and shown in equation (8.5.4), when noise is white in character, the effective noise level can be reduced by limiting the bandwidth of detection to that just necessary to include the range of frequencies occurring in the signal, thus reducing the noise power, but not the signal. Phase sensitive detection overcomes the practical problems. It enables extremely narrow bandwidth detection with a bandwidth in the range of 10^{-3} Hz with stabilities measured in parts per million.

All lock-in amplifiers, whether analogue or digital, rely on the concept of phase sensitive detection for their operation. Phase sensitive detection requires the measurand to be modulated at a discrete frequency that is about ten times higher than the highest frequency required from the measurand. After initial amplification, and possibly wide-band filtering, the modulated signal is synchronously detected using the reference signal to form the product. The basic phase sensitive detector can be regarded as a multiplier plus low-pass filter.

A basic schematic of the lock-in operation is shown in Figure 9.6. Generally the input has a broadband amplifier for the low level input signal. The amplified signal is then sent to a "multiplier". In the simplest approximation the multiplier can multiply the input signal by 0 or 1. A realization of that is shown in Figure 9.6 which is a simple switch which is operated by a waveform derived from the reference frequency that modulates the experiment. When the reference signal closes the switch circuit, the output of the amplifier is connected directly to a low pass filter consisting of a resistor (R) and a capacitor (C). When the reference signal is such that the switch is open no connection is made.

Rectification of the signal occurs when the waveform controlling the switch is exactly in phase with the ac signal at the input to the switch. The circuit thus operates as a "synchronous rectifier" and the noise associated with the signal that is at random phase with the lock-in passes un-rectified to the low pass filter. Beyond this filter it is smoothed or averaged to its mean value of zero. The out put filter shown in the figure, as a *RC* filter, is the simplest realization. In real lock-in amplifiers one has digital filters, described in the section before, whose roll-off can be adjusted to make it a very sharp filter (like 120 db/octave). If the roll-off occurs at a frequency of ω_c (like the 3 db point of a first order low pass filter) the effective bandwidth of the signal detection is limited to only ω_c. This can be adjusted by the user of the lock-in amplifier. However, it need be ensured that the response required in the experiment is slower than $1/\omega_c$. The operation of the device has all the features of a tuned amplifier followed by a rectifier. However, it is much more than a tuned amplifier. The center frequency of the filter is locked (hence lock-in) onto the reference frequency. This feature makes sure that the signal is locked to the pass band of the filter.

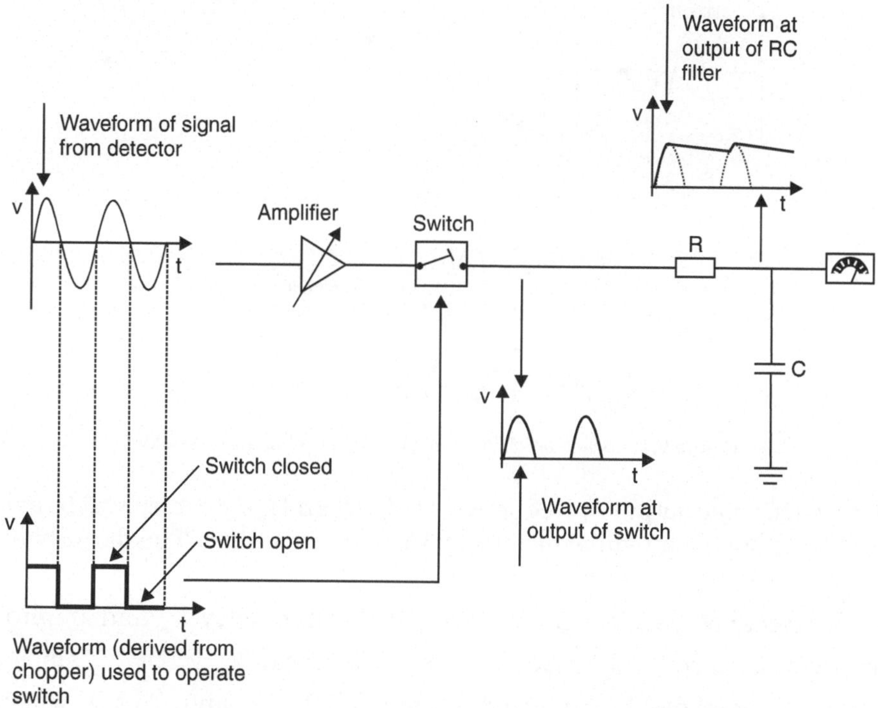

Figure 9.6: Basic detection scheme of a lock-in amplifier

So far we have considered the phase sensitive demodulator as a single pole switch which is capable of synchronous rectification of only half of the signal. For a signal recovery device such a waste of signal information would be unacceptable. Commercial lock-in amplifiers therefore include a full wave synchronous demodulator that usually works by using inverting and non-inverting amplifiers to produce anti-phase versions of the signal (i.e. 180° out of phase). This is shown in the cartoon in Figure 9.7. This ensures a better noise rejection.

How does the lock-in amplifier convert the signal to a dc? In a "lock-in" the "multiplier" multiplies the signal of frequency component f_s with the reference at frequency f_r. Assuming that both the signal and reference waveforms are sinusoidal then the output of the multiplier will contain components at frequencies of $f_s + f_r$ and $f_s - f_r$ where f_s and f_r are the frequencies of the signal and reference waveforms respectively. If $f_s = f_r$, as is the case where the reference waveform is derived from the device which is modulating the signal, then there will be an output at 0 Hz i.e. dc. Any other component in the signal e.g. a noise component at a frequency of f_n

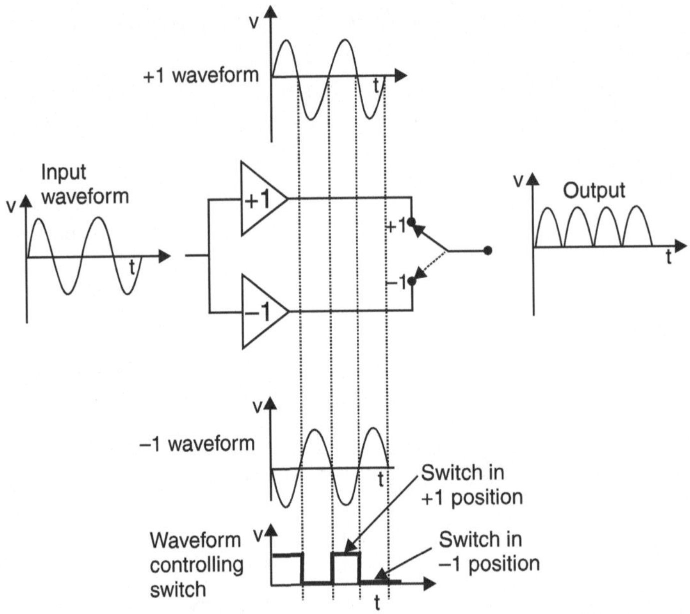

Figure 9.7: Lock-in amplifier as a full wave synchronous detector

will give rise to an ac output at frequencies of $f_n - f_r$ and $f_n + f_r$ which will be smoothed or averaged to the mean value of zero by the low pass filter. The discussion above can be described mathematically below.

Let the signal be a sinusoidal wave of circular frequency ω_s and amplitude V_{os} and phase (with respect to some arbitrary reference) ϕ_s, $V_s(t) = V_{os}\sin(\omega_s t + \phi_s)$. The reference signal also is a sinusoidal signal $V_r(t) = V_{or}\sin(\omega_r t + \phi_r)$, where V_{or} is the amplitued, ϕ_r is the phase and ω_r is the frequency. These two signals when multiplied gives the signal V_m as:

$$V_m = V_{or}\sin(\omega_r t + \phi_r).V_{os}\sin(\omega_s t + \phi_s)$$

$$= (1/2)V_{or}V_{os}\cos(\{\omega_s - \omega_r\}t + \{\phi_s - \phi_r\}) + (1/2)V_{or}V_{os}\cos(\{\omega_s + \omega_r\}t + \{\phi_s + \phi_r\}) \quad (9.4.1)$$

Since the signal is modulated at the same frequency as the reference we have

$$V_m = (1/2)V_{or}V_{os}\cos(\phi_s - \phi_r) + (1/2)V_{or}V_{os}\cos(\{\omega_s + \omega_r\}t + \{\phi_s + \phi_r\}) \quad (9.4.2)$$

The output filter being a low-pass one will filter out the ac signal of sum frequency and the output will be a dc:

$$V_m = (1/2)V_{or}V_{os}\cos(\phi_s - \phi_r) \quad (9.4.3)$$

From equation (9.4.3) we can see that the out-put signal from the lock-in will be maximum when the reference and signal are in phase (or $(\phi_s - \phi_r) = 2n\pi$) and will go to zero for $(\phi_s - \phi_r) = (2n+1)\pi/2$.

Let us consider an example to illustrate the extent of noise reduction by a lock in amplifier and the extent of *S/N* achievable in presence of extreme noisy low level signal. Suppose the signal is a 5 nV sine wave at 10 kHz. The signal will be pre-amplified by a low noise amplifier. A good low noise amplifier will have about $5\,\text{nV}/\sqrt{\text{Hz}}$ of input noise. If the amplifier bandwidth is 100 kHz and the gain is 1000, then the output will be a signal of 5 μV. This will also be accompanied by a broad-band noise. This on a bandwidth of 100 KHz and gain of 1000 will be ≈1.6 mV. Presence of this noise will make the 5 μV signal almost impossible to detect in spite of a high amplification (~10^3) by a low noise amplifier.

In a lock in amplifier the bandwidth can be limited to 10^{-2} Hz. (This for a 10 kHz signal will mean a band pass filter of $Q = 10^6$!). In this case, the noise in the detection bandwidth will be only 0.5 μV while the signal is still 5 μV. The signal to noise ratio is now 10 and an accurate measurement is possible.

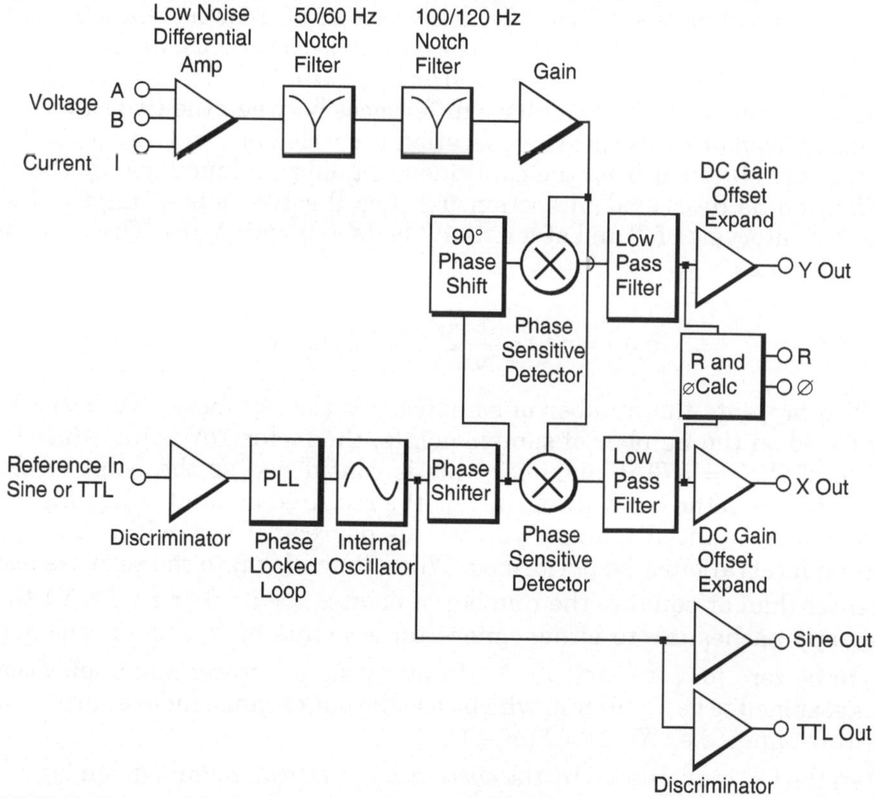

Figure 9.8: Block diagram of the functional blocks of a commercial lock-in amplifier SR830 of Stanford Research [5]. The block diagram is taken from the manual of the lock-in amplifier

The Figure 9.8 is the functional block diagram of a modern commercial lock in amplifier. In these equipments the complete operation is done by digital techniques. The diagram shown above is for a two-phase model that can simultaneously measure signal that is in phase and in quadrature with the reference. This allows mea-surement of both the amplitude and phase of the input signal.

5. CROSS-CORRELATION TECHNIQUES FOR S/N ENHANCEMENT

For detection of signals like those seen in ultrasonics, time resolved reflectometry, radar etc where a propagated signal is detected after reflection, or complex signals like those found in electrocardiogram, cross-correlation technique is a very powerful technique. In this technique as the name cross-correlation suggests, the experimental signal is compared with the desired signal to detect how much of the desired signal is contained within the detected signal. CC techniques are widely used in 2D digital image processing to recover an image from a noisy environment.

The Cross-Correlation (CC) function can be used to measure the similarities between two signals. A good method of measuring the similarity between two wave-forms is to multiply them together, coordinate by coordinate and to add the products over the duration of the wave forms. Thus the process of cross-correlation, essentially involves the computation of the sum of the products of corresponding pairs of points of two signals, within a specified time frame, or window. The process also takes into consideration any potential phase differences between the two signals via the inclusion of a lead or lag term. The operation is carried out on two sets of digitized data which are obtained from the continuous signals (or functions) by the process of digitization as described in section 9.2. Let the two sets of digitized signals (digitized at intervals of T and of length N) be $\{X_1(i)\}$ and $\{X_2(i)\}$. Then the formula for CC is given as:

$$R_{12}(j = nT) = \frac{1}{N}\sum_{i=0}^{N-1} X_1(i)X_2(i + j) \tag{9.5.1}$$

where N represents the number of samples, j is the lag factor. To normalize the results based on the number of sample points, the factor $1/N$ is introduced. When the value of $R_{12}(j = nT)$ is maximal, this is considered as the point of maximal similarity between the two waveforms. As the required amount of lagging between the two signals is initially unknown, various degrees of lags within the specified correlation interval must be performed. When the index into the series is less than 0 or greater than or equal to the number of points. ($i + j < 0$ or $i + j \geq N$) the most common approaches are to either ignore these points or to assume the series X_1 and X_2 to be zero for $i < 0$ and $i \geq N$. In many signal processing applications the series is assumed to be circular in which case the out of range indexes are "wrapped" back within range, i.e.: $X(-1) = X(N - 1)$.

Often the CC is evaluated by the *correlation coefficients* for a given lag j defined below:

$$\tilde{R}_{12}(j = nT) = \frac{\sum_{i=0}^{N-1} \{X_1(i) - \overline{X}_1\}\{X_2(i + j) - \overline{X}_2\}}{\sqrt{\sum_{i=0}^{N-1} \{X_1(i) - \overline{X}_1\}^2 \{X_2(i) - \overline{X}_2\}^2}} \tag{9.5.2}$$

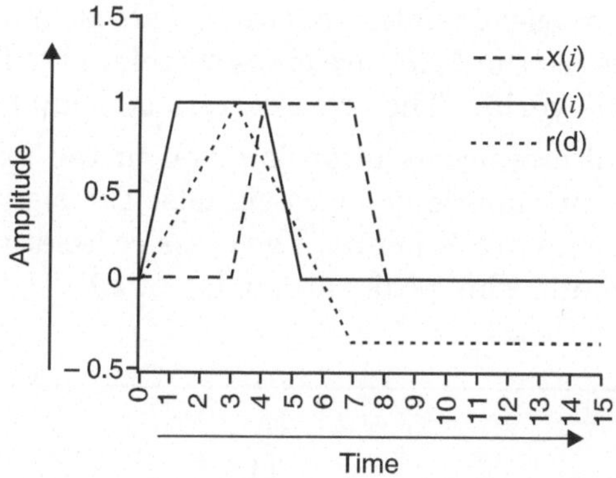

Figure 9.9: Cross-correlation of two square signals as a function of delay. The value of the cross-correlation reaches a maximum at the delay where the overlap is maximum.
(adopted from http://astronomy.swin.edu.au/~pbourke/analysis/correlate/)

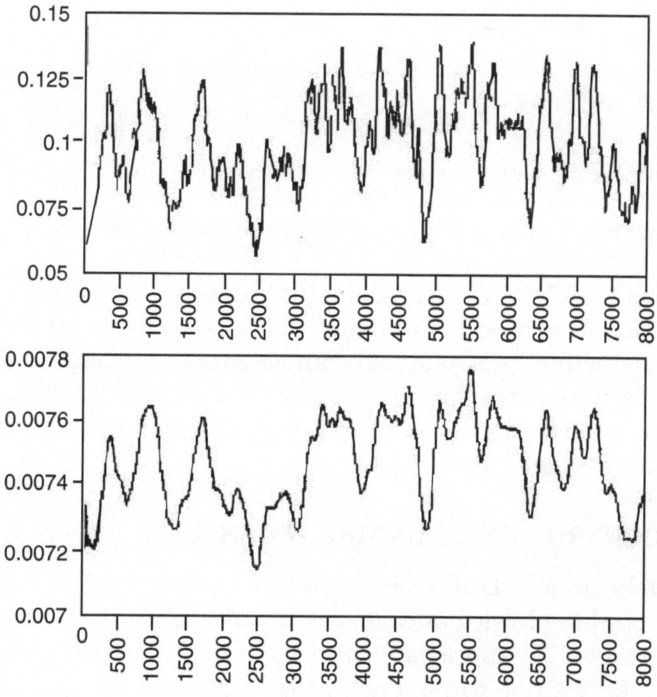

Figure 9.10(*a*): Time series of two random signals

This definition follows the standard statistical definition of correlation coefficients for the zero lag (j = 0). As before the maximum overlap of the two signals is for a given lag which is found out by finding the peak in $\widetilde{R}_{12}(j = nT)$ or, R_{12} ($j = nT$) as n is varied.

An example of the cross-correlation of two signals is shown in Figure 9.9. Here two rectangular signals ($x(i)$ and $y(i)$) are cross-correlated to find the delay at which they have the maximal overlap. The cross-correlated signal is shown as $r(d)$.

Another example of cross-correlation is given in the example in Figure 9.10 where both the signals are aperiodic noisy signals. The signals x and y are shown as time series with time up to 8 K points. The cross-correlation has been calculated up to a delay of 4 K points. This peaks at a delay of 40.

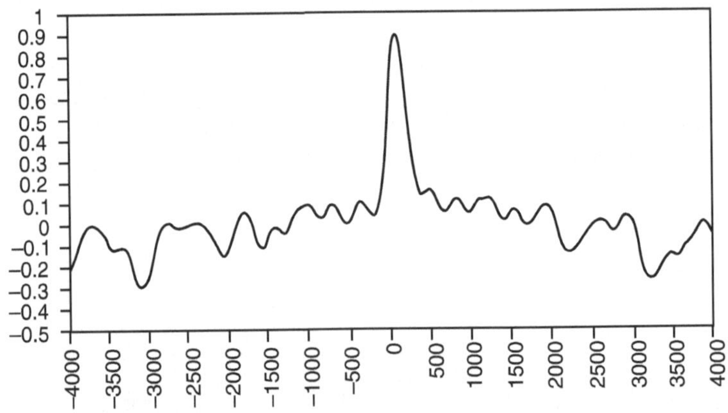

Figure 9.10(*b*): Cross-correlation function of the two time series shown in Figure 9.10(*a*). The value of the cross-correlation reaches a maximum at the delay where the overlap is maximum. (adopted from http://astronomy.swin.edu.au/~pbourke/analysis/correlate/)

Using a known template, one can detect its presence in the signal calculating the cross-correlation as a function of delay. This is an interesting application of CC technique in detecting known signal components in a random time series. An example of this is multi-component based cross-correlation beat detection in electrocardiogram analysis. This article is available from: http://www.biomedical-engineering-online.com/content/3/1/26/.

6. FEED BACK CONTROL USING DIGITAL TECHNIQUES

An important application of DSP techniques is its use in control loops. Feed back control is widely used in physical measurements. A widely used feed back loop is a temperature controller. To appreciate the role of digital signal processing in the feed back loop we take a real example of a *temperature controller* first, discuss its operation and then discuss how digital signal processing can be incorporated. A basic discussion of temperature control can be found in [6].

In Figure 9.11 we show a schematic of a cryostat. The cryostat contains a vacuum chamber that has a thermally controlled block of copper whose temperature is measured by a thermometer. A heater supplies electrical power to heat the block by resistive heating above the temperature of the coolant which in this particular example is a liquid cryogen like liquid He (bath temperature 4.2 K) or liquid nitrogen (bath temperature 77 K). Or the block can be mounted on a cold finger of a cryocooler. The temperature-controlled block will have the experiment on it. The temperature of the block is controlled by a feed back loop. To understand its operation we can have a thermal model shown in Figure 9.12. In this model a heater, of thermal capacity C_h at a temperature T_h, is attached to the thermally controlled block (thermal capacity C_o) by a link of thermal resistance R_h. A power ($\dot{Q}_m$) is supplied to the heater to maintain the temperature (T_0) of the block. The block is connected to the cryogen bath (or cold finger) at temperature T_c by a thermal link of resistance R_o. The highest temperature that the block can go to is determined by the thermal link R_o and is given by the static heat balance equation:

$$(T_0 - T_c) = \dot{Q}_m . R_0 \tag{9.6.1}$$

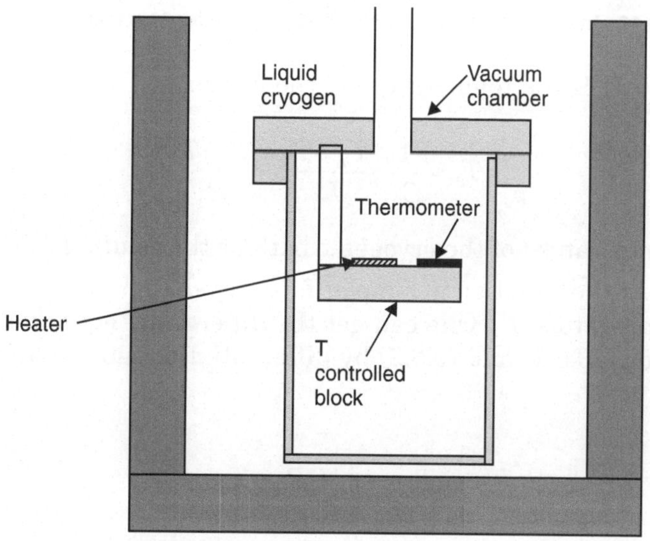

Figure 9.11: Schematic of a cryostat with a temperature controlled block

Here $\dot{Q}_m$ is the maximum power that can be supplied to the heater by the power supply. R_0 can be determined by passing known amount of power $\dot{Q}$ to the heater and then finding T_o. Generally $R_h << R_0$ and $C_0 >> C_h$ but for complete modeling it is important to keep these terms. The differential equations for T_h and T_0 can be written as:

$$\dot{T}_h = \frac{1}{C_h}\left\{\dot{Q} - \frac{T_h - T_0}{R_h}\right\} \tag{9.6.2}$$

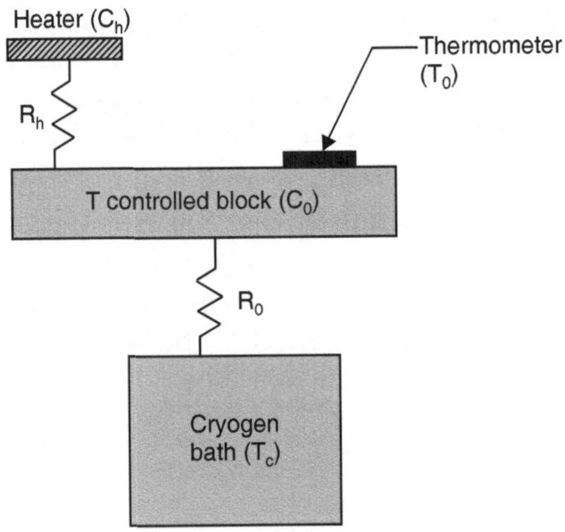

Figure 9.12: The thermal model used for designing the feed back temperature control

In the above T_h is the temperature of the heater.

$$\dot{T_0} = \frac{1}{C_0}\left\{\frac{T_h - T_0}{R_h} - \frac{T_0 - T_c}{R_0}\right\} \tag{9.6.3}$$

T_c is the temperature of the cryogenic bath or the coldfinger.

We can recast the equation in simpler forms by defining T_h and T_0 with respect to the bath temperature T_c. One can get the differential equation for T_0 from the coupled equations (9.6.1) and (9.6.2) by differentiation and elimination of T_h and its derivative:

$$\ddot{T_0} + 2\gamma_0\omega_0 \dot{T_0} + \omega_0^2 T_0 = \omega_0^2\left\{\dot{Q}R_0 + T_c\right\} \tag{9.6.4}$$

In this equation the system parameters enter through dimensionless damping parameter γ_0 and the natural frequency ω_0. They are defined through the thermal capacities and the thermal resistances as given below:

$$\omega_0 = (C_h R_h C_0 R_0)^{-1/2}, \gamma_0 = \left(\frac{\omega_0}{2}\right)(C_0 R_0 + C_h(R_0 + R_h)) \tag{9.6.5}$$

The task of the feed back loop is to set a value for $T_0 = T^*$ then adjust $\dot{Q}$ so that T_0 is kept at that value. The simplest from of control is the proportional control where the heat input is proportional to the deviation from the set value so that

$$\dot{Q} = \frac{g_p}{R_0}(T^* - T_0) \tag{9.6.6}$$

where g_p is a dimensionless time constant. With such an input for the power $\dot{Q}$ the differential equation (9.6.4) gets modified with renormalized constants:

$$\ddot{T}_0 + 2\gamma_p\omega_p \dot{T}_0 + \omega_p^2 T_0 = \omega_p^2 \{g_p T^* + T_c\} \tag{9.6.7}$$

where the renormalized constants are defined as:

$$\gamma_p = \gamma_0(1+g_p)^{-1/2} \quad \text{and} \quad \omega_p = \omega_0(1+g_p)^{1/2}$$

The effect of the proportional controller is to speed up response by reducing the damping constant but this actually makes the system unstable. Apart from the instability problem a proportional controller suffers from zero error i.e., the temperature stabilizes at a point different from T^* in order to have a finite $\dot{Q}$ (see eqn. 9.6.6, when $T_0 \to T^*$, $\dot{Q} \to 0$). The problem of overshoot and the stability is improved by adding an additional damping term so that the heat supplied is now given by:

$$\dot{Q} = \frac{g_p}{R_0}(T^* - T_0) + \frac{g_d}{\omega_0 R_0}(T^* - \dot{T}_0) \tag{9.6.8}$$

The dimensionless gain constant g_d can be adjusted independently so that system can reach critical damping condition given by $\gamma_0 + g_d/2 = \sqrt{(1+g_p)}$. The zero error is removed by adding an integral term. This makes a PID (proportional-integral-differential) controller. In this case the total heat balance is given as:

$$\dot{Q} = \frac{g_p}{R_0}(T^* - T_0) + \frac{g_d}{\omega_0 R_0}(\dot{T}^* - \dot{T}_0) + \frac{g_i\omega_0}{R_0}\int (T^* - T_0)dt \tag{9.6.9}$$

The presence of the integral term guarantees a finite $\dot{Q}$ when $T_0 \to T^*$ and thereby reduces the zero error. However the response of the system becomes more complex and the stability is determined by the inequality $1 + g_p \geq \frac{g_i}{g_d + 2\gamma_0}\gamma_0$. This determines the parameter space within which to set the system. In analogue controllers op-amps are used to implement equation (9.6.8). In modern controllers one can use DSP chips to implement the PID controllers. In many digital PID temperature controllers one can store in a table the parameters needed for the PID loop for its optimum operation and the controller sets these parameters automatically, instead of setting it by hand.

The *Zeigler Nichols Methods* [7] have been well used to choose the best parameter for a feed back loop. There are two methods to determine the ultimate parameters used. The first method, which is a frequency response method, uses the stability of the feed back loop. This method sets the system to the critically damped mode. In this method the integral gain g_i and the differential gain g_d are set equal to 0. The set point value is entered in the controller and the feed back loop is enabled by putting a low value for the proportional gain g_p. The gain g_p is slowly increased until the system starts to oscillate. The gain g_c and the oscillation period T_c are noted. The optimum parameters are then obtained from the critical gain g_c and the period T_c.

In the second method, which is a transient response method, the system is allowed to attain a higher temperature from a set equilibrium value and the time delay L and the steepest slope R are deduced. The PID values are obtained from L and R.

In more sophisticated auto-tuning PID controllers the optimum parameters are generated automatically. A related issue in PID control that becomes important in auto-tuning is the effect of system noise in the feed-back control. We do not go into detailed discussion on this issue as this is beyond the scope of the book. The commercial temperature controllers perform auto-tuning by applying a well-defined synthesized waveform to the heater output and analyzing the resulting changes in the temperature. This response is used to develop a process model as given in equations (9.6.4) and (9.6.8) and then a PID solution is generated. Due to the presence of feed-back loop noise there is a range of PID combinations that will provide accurate control for a given process. Further, process modeling is a statistical method that is affected by noise and system non-linearity.

For a general PID controller output voltage $V(t)$ in analogue form is related to the error signal $E(t)$ by the equation:

$$V(t) = PE(t) + D\frac{dE}{dt} + I\int_0^t E(t)dt \tag{9.6.10}$$

where the error signal at any time t is $E(t) = K(\eta_s - \eta(t))$, η_s is the set value of the controlled quantity $\eta(t)$. P, I, D and K are constants. In a digital feed-back loop equation, the inputs are the digitized voltage and the integrals and derivatives are appropriately redefined. Equation (9.6.10) in digitized form is given as:

$$V_n = PE_n + D\frac{E_n - E_{n-1}}{\delta t} + I\delta t\sum_{i=1}^{n} E_i \tag{9.6.11}$$

In equation (9.6.11), V_n is the nth output of the heater voltage and it can be initialized as $V_n = 0$ and $E_n = 0$ for $n = 0$ and $\delta t = 1/f_s$, f_s being the sampling frequency. In equation (9.6.11) the integral of equation (9.6.10) is replaced by the sum and the derivative by the first difference of consecutive signals.

A simple DSP based practical temperature controller is shown in Figure 9.13. The controller is built around a Texas Instrument DSP and it talks to a PC by an

RS 232 interface. The PC runs the DSP card using its own programme language supplied by the manufacturer and is based on a back bone programme running in *C*. The voltage across the thermometer (which is current biased like a diode or resistance thermometer) is amplified and then read by the AD card of the DSP board. It is 16-bit card with highest digitization accuracy (in 1.5 V full scale) of 45.776 µV. The voltage is compared to the set voltage (which is determined by the temperature to be set by the calibration) and the error voltage is calculated. The DSP card then calculates the out put voltage to be sent to the heater using equation (9.6.10), and sends it to the DA card. The calculation is made after each sampling of the input sensor voltage by the AD card. Typically the response of the thermal system is much slower than the digitization frequency of the ADC. The voltage out put from the DA card needs to be current boosted by a power amplifier (often a Darlington pair would suffice) and then sent to the heater. The controlled sensor voltage is sent to the main PC and is read along with other experimental parameters. The controller can control to better than 10 mK close to room temperature and better than 1 mK down to liquid nitrogen temperature.

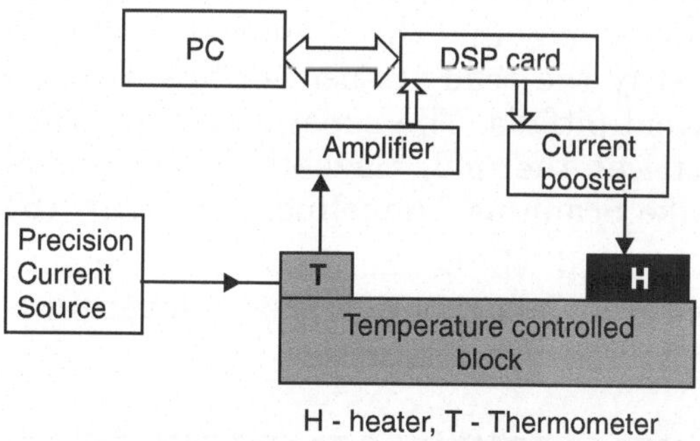

Figure 9.13: A DSP based digital temperature controller

REFERENCES

1. DeFatta D., Lucas J.G. and Hodgkiss W.S., "Digital Signal Processing: A System Design Approach", John Wiley & Sons, (1988).
2. There are a number of vendors for such cards for instance National Instruments Corporation, 11500 N Mopac Expway, Austin, TX 78759, USA, www.ni.com and ADVANTECH, No. 1, Alley 20, Lane 26, Rueiguang Road, Neihu District, Taipei 114, Taiwan, www.advantech.com.tw.
3. Nyquist H., Trans. AIEE 47, 617 (1928). Shanon C.E., Proc. Inst of Radio Engg. 37, 10 (1949).
4. Hamming, R., "Digital Filters", Prentice Hall, Englewood Cliffs, N.J., (1977).
5. Stanford Research Systems, Inc. 1290-D Reamwood Ave, Sunnyvale, CA 94089, USA, www.srsys.com.
6. Dogan Ibrahim, "Microcontroller Based Temperature Monitoring and Control", Newnes (2002).
7. Vance J. and Van Doren, "Basics of Proportional-Integral-Derivative Control", Control Engineering (March 1, 1998).

10 CHAPTER

Solid State Measurements at Low Temperatures

1. INTRODUCTION

In this chapter we will take examples of certain low temperature measurement techniques. The chapter is mainly written for those who would like to integrate such experiments by using commercially available measuring equipments and adding innovation to it. Now-a-days there are ready-made complete systems available that can measure some of these properties. Those who would like to utilize such ready-to-use systems may find the chapter helpful because it can give them an insight and demystify the ready-to-use set-ups. Often a basic knowledge of measurements can avoid pitfalls. The experiments described in this chapter are low temperature electrical, thermal, magnetic measurements and such spatially resolved techniques like Scanning Tunneling Microscopy (STM).

The experiments covered will be mostly over the temperature range 0.3 K to 500 K. For most measurements this is a good temperature range that can give useful information.

2. HIGH PRECISION MEASUREMENT OF LOW RESISTANCES

In general, dc resistance measurement is the most convenient technique for resistances that are moderately high (>100 ohm). However, when the resistance is low, the voltage to be measured also becomes low and in this situation one can use a low frequency (~1 KHz) a.c. measurement using a lock-in amplifier that allows phase-sensitive detection and significant improvement in the signal-to-noise ratio. In particular, at low temperatures where one cannot dissipate much power, high precision measurements will need special care. We should also note that in an ordinary laboratory the power line at 50 Hz will contribute approximately 10^{-11}–10^{-12} V^2/Hz noise spectral power within a band width of say 10 Hz unless a special shielded room is used. This would mean an rms noise voltage of around 0.3–1 μV. We call this the contribution of the environment and it is represented by an rms voltage v_E. Additional noise will come from the thermal noise, which, integrated over the experimental band width Δf_{exp}, will contribute an rms noise $v_{thermal} \sim (4k_B TR\Delta f_{exp})^{1/2}$. These two will determine the noise floor. So any measurement that we do will be limited in its precision at least by these factors. In addition, there will be other noise contributions coming from the amplifier and such electronics. Within the bandwidth of detection we get an rms voltage representing it. This is given as v_A. This is generally given by the manufacturer in the manual of the instrument. It is a good practice to read the manuals of good manufacturers because they often contain information that is not available elsewhere and they can make an experimentalist use the equipment in the best possible manner. Without knowledge of these in the resistivity measurement we may use the costliest of electronics but will not be able to take advantage of it fully. In general the noise voltage (expressed as rms noise) is given as total contribution of the three and can be expressed as:

$$v_{\text{noise}} = (v_E^2 + v_A^2 + v_{\text{thermal}}^2)^{1/2} \quad (10.2.1)$$

However, the minimum voltage that we can measure in an experiment is determined not by v_{noise} alone. This will also be determined by the resolution of the voltmeter/DMM and we represent that as v_R. If $v_R >> v_{noise}$ the experiment will be resolution limited and if $v_{noise} >> v_R$ it will be noise limited. We can combine these two and find out the minimum detectable voltage in a set up by:

$$v_D = (v_R^2 + v_{noise}^2)^{1/2} \quad (10.2.2)$$

The measurement of resistance by an apparatus will only be possible if $v_D << v_{\max}$, where $v_{\max}$ is the maximum signal that we can get and is mostly determined by the maximum power that can be dissipated to determine the resistance before heating it up. We may follow the recipe to determine $v_{\max}$ from the Table 10.1 below.

In the Table 10.1 below we also give a check list of things that will help the experimenter to decide on how to make measurements. This has been done in the context of resistance measurements but it is of general use.

Table 10.1: The factors one should check to optimize the resistance measurements

Task	*What determines it*	*How to estimate it*
1. Maximum power that can be dissipated ($Q_{\max}$)	Thermal "grounding" (resistance) to base.	This can be estimated from the approximate surface area over which the sample makes contact with the base, the thermal conductivity of the material that makes the thermal contact, boundary resistance and the lead thermal conductance. If this is not known then make a dummy run to find out how much power really can be used. This depends on temperature.
2. Maximum measuring current that can be used $I_{\max}$	$Q_{\max}$	$I_{\max} = (Q_{\max}/R_t)^{1/2}$ where R_t is the two probe resistance of the sample i.e., the resistance between the current leads including contact resistance.
3. Maximum signal that can be used	$v_{\max}$	$v_{\max} = I_{\max} \cdot R_s$, R_s is the 4-probe resistance.
4. Noise voltage	v_{noise}	This contains contribution from three main sources: environment (mainly 50 Hz), amplifier noise and thermal noise (Nyquist noise).
5. DMM resolution	v_R	Manufacturerer's manual
6. Minimum detectable voltage	v_D	$v_D = (v_{\text{noise}}^2 + v_R^2)^{1/2}$

As stated before, measurement of the resistance with a given apparatus is possible only when v_{max} is substantially larger than v_D. The method to be followed (d.c. vs a.c. etc.), what voltmeter to be used and what excitation voltage to be used (which is determined by v_{max}) can all be intelligently decided. Generally the problem is most severe at low temperatures because Q_{max} is limited. As the temperature increases Q_{max} increases substantially and this allows one to use more v_{max}. The problem is more compounded when a low resistance need be measured. We can illustrate this by the following examples. Let us say we have to measure 100 ohm and 1 milliohm resistances both with a moderate resolution of 1 part in 10^4 in a four-probe configuration. In case of R_s = 100 ohm it will mean smallest resistance that we need to measure $\delta R \approx 10^2 \times 10^{-4} = 10^{-2}$ ohm and that in case of R_s = 1 mohm is $\delta R \approx 10^{-3} \times 10^{-4} = 10^{-7}$ ohm. Suppose the experiment is being carried out in the vicinity of 1 K. The typical Q_{max} is often in the vicinity of few hundreds of nW. Let us assume that we have an estimated $Q_{max} \approx 500$ nW. If we assume the samples have good contacts and the voltage leads are not too far away from the current leads R_t, the two probe resistance will be only few times the 4-probe resistance R_s.

For example let R_t for the 100 ohm sample be 200 ohm and that for the 1 milliohm sample be 2 milliohm. The v_{max} allowable for the 100 ohm resistance is 5 mV with $I_{max} \approx 50$ μA. For achieving a resolution of $\delta R \approx 10^{-2}$ ohm one needs $v_D \leq 0.5$ μV. This is achievable with d.c. DMM/DVM with 5½ digit resolution ($v_R \approx$ 0.1 μV), a good level of shielding for the 50 Hz signal (so that $v_R \leq 0.3$ μV) and moderate bandwidth of 1 KHz to limit the thermal noise, which in the vicinity of 1 K for 100 Ohm is ≈ 75 pV/Hz$^{1/2}$. (At 300 K thermal noise is 1.5 nV/Hz$^{1/2}$ for the same 100 Ohm resistor). For the 100 Ohm resistor a simple lock-in arrangement shown in Figure 10.1 will be a better solution.

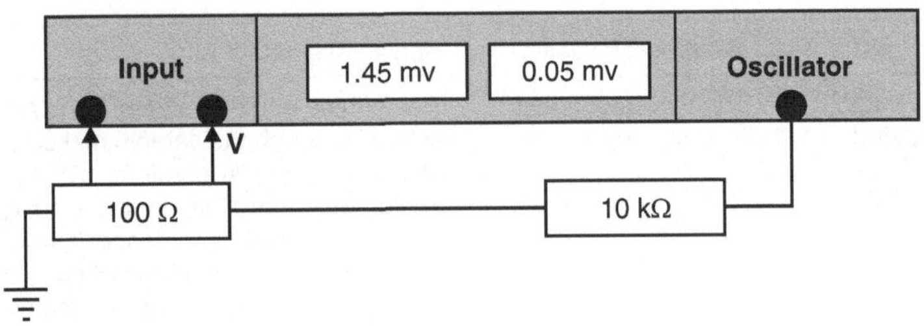

Figure 10.1: Measurement of resistance by a lock-in

This is because a lock-in amplifier allows the effective bandwidth to be reduced causing a severe reduction in v_{noise} as described in chapter 9. The 10 Kohm drop resistor allows the lock-in out put to become an effective current source. The voltage drop across this resistor can be measured to get the correct current. The drop resistor cannot be made very high because it will cross the voltage compliance of the lock-in (usually 10 V). Also for R > 100 Kohm the R becomes comparable to $1/\omega C_p$ where ω = measuring frequency and C_p is the parasitic capacitances due to coaxial cables. This will introduce a good amount of out of phase signal that will change as the temperature changes. (For instance a moderate $C_p \approx 100\ pf$ and a measuring frequency of 1 KHz will give $1/\omega C_p \approx 1.6$ Megohm. This implies a phase error of 3.5°).

In case of the 1 milliOhm resistor the numbers will take us to a different regime of measurement. The v_{max} allowable is 16 μV with $I_{max} \approx 16$ mA. For achieving a resolution of $\delta R \approx 10^{-7}$ ohm one needs $v_D \leq 1.58$ nV. This is almost impossible to attain in a d.c. measurement and even in lock-in based conventional a.c. method shown in Figure 10.1. For this range one needs a bridge configuration that allows a much larger magnification. Below we describe a realization of such a technique shown in Figure 10.2 which can give 10 ppm resolution in resistance measurements in the range 10 Ohm down to few milli-Ohm.

This is a high precision four-probe *ac bridge* (20 Hz) with a PC based automatic data acquisition program. The same measuring current is made to pass through the sample (R_s) and a standard resistance (R_{std}). The voltage developed across R_s is amplified by a low noise transformer by a factor G and balanced off against a fraction of the voltage developed across the standard. (For low noise shielded transformers G can be ≈ 100). The fraction of the voltage across the standard is chosen using a Kelvin-Varley voltage divider. The voltage at the output of the bridge is v_{out} and is given as:

$$v_{out} = I(fR_{std} - GR_S) \qquad (10.2.3)$$

G is the gain from the transformer (1 : 100 low-noise shielded). Low noise transformer also impedance matches the low impedance sample with the high input impedance of the lock-in amplifier which reduces the noise. In the balanced condition $v_{out} = 0$ and $f = f_{bal}$ so that $R_S = \dfrac{f_{bal}R_{std}}{G}$.

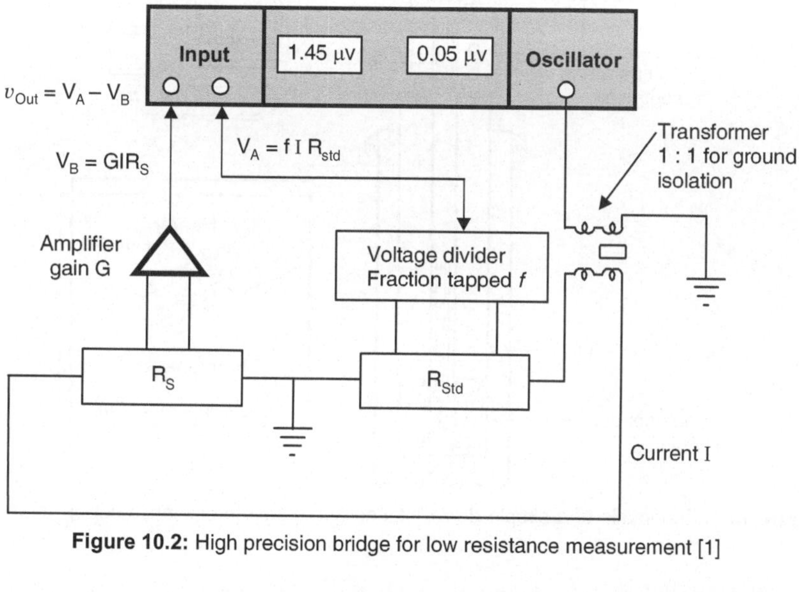

Figure 10.2: High precision bridge for low resistance measurement [1]

As the resistance (i.e., R_s) changes (with change of temperature or magnetic field as in case of magneto-resistance measurements) the bridge becomes off-balanced and this signal Δv_{out} is measured by the lock-in. Δv_{out} is recorded by the computer. Δv_{out} is given by:

$$\Delta v_{out} = If_{bal}R_{std} - IG(R_S + \Delta R_S) = IG\Delta R_S \tag{10.2.4}$$

ΔR_S is the change in resistance. In general if the resistance change $\Delta R_S << R_{std}$, the current I is constant within the ratio $\Delta R_S/R_{std}$. In this case R_{std} being much larger than R_S makes the circuit a constant current circuit. For most measurements small variation of I is of no consequence. However, for very precise measurement it is better to track even a small change in I. This can be done by measuring the ac voltage across the standard resistance with a DMM or another lock-in. The change in resistance

$$\Delta R_S = \frac{\Delta v_{out}}{IG} \tag{10.2.5}$$

In this case, a high gain low noise amplifier allows one to measure a small voltage. For instance a gain of G = 100 would imply a voltage of 100 nV.

In general for dc resistance measurements one uses a current source and a Digital Voltmeter (DVM) or multi-meter (DMM) that is capable of measuring desired

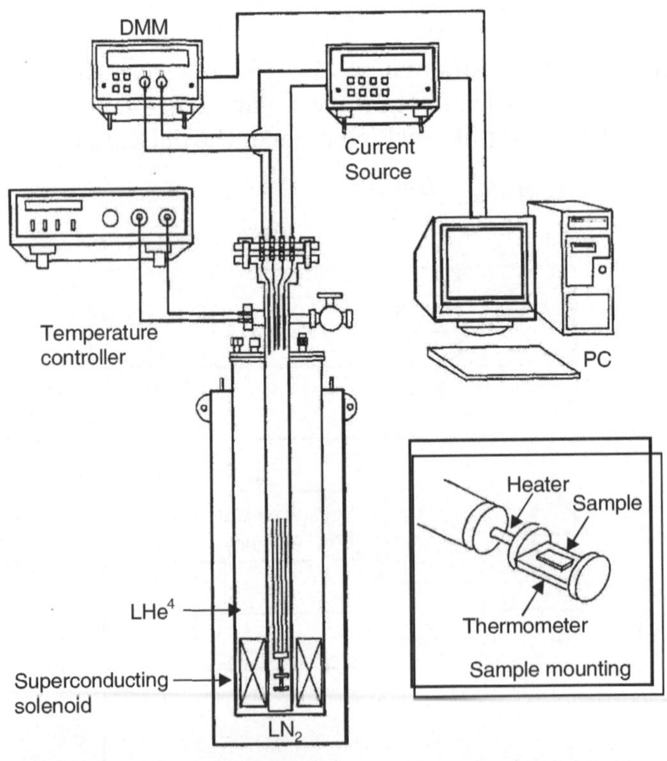

Figure 10.3: Schematic of a simple dc resistance and magnetoresistance set-up (from [2])

voltages. For measurements of high resistance one uses an electrometer. Often if one desires to measure more than one sample at a time use of a scanner card is desirable. What DMM, DVM or scanner card to use will depend on the voltage to be measured. For a detailed description of these issues the reader is referred to the booklet "Low Level Measurement" published by the electronic test and measurement manufacturer Keithley®. In particular this book discusses in detail the high impedance measurements. The same resistance measurement set-up when used in a magnetic field will give us the *magneto-resistance*. A simple set-up is given schematically in Figure 10.3.

3. MEASUREMENTS OF SEEBECK COEFFICIENT (THERMOELECTRIC POWER)

In most solid state physics measurements, when we talk of Thermoelectric power (TEP) we refer to one of the coefficients, namely the Seebeck coefficient (S), although we can relate S to other coefficients. The TEP is an extremely sensitive property whose study as a function of temperature would give valuable information on the electronic structure of the system under study. A simple physical interpretation of TEP can be found in Solid State Physics books [3, 4].

The Seebeck coefficient (S) is defined as follows:

$$S = Lt_{\Delta T \to 0}\left(\frac{\Delta E}{\Delta T}\right) \tag{10.3.1}$$

where ΔE is the differential emf developed across the sample for a temperature difference ΔT. From the definition we can immediately see that S can be measured using the differential technique wherein we create a temperature gradient ΔT across the sample and measure the differential emf ΔE developed between the hot and cold ends of the thermocouple formed with a reference material. The difficulty in the measurement of the TEP in most cases is its small magnitude especially if the sample is a metallic one. Typically $S \propto n_{\text{carrier}}^{-1}$, where n_{carrier} is the carrier concentration of the solid in question. For many metallic solids the TEP decreases as T decreases and may go below 1 μV/K which makes its measurement so much more difficult. As an example suppose we want to measure S = 1 μV/K at 10 K with a modest accuracy and precision of 10%. The maximum ΔT should not extend beyond 5% of T and is thus ≈ 0.5 K. Expected $\Delta E \approx 0.5\,\mu$V. To have an accuracy and precision of $\pm 5\%$ we should have $|v_D| \le 25$ nV. At this level of measurement the small inhomogeneity in Cu leads may also matter. For semi-conducting samples, whose resistivity increases with decreasing temperatures, the value of S also increases. But the high resistance of the sample makes the noise increase $|v_D|$ and it may increase to such a high value that measurements are difficult. Also a high sample resistance will necessitate the use of a high input impedance voltmeter or electrometer, which also adds its own noise.

How do we determine the sign of the thermopower? This question is often of importance. With the cold end of the sample connected to the positive of the measuring instrument the sign of the voltage developed gives the sign of the thermo-emf difference between the sample and the reference material. There are many descriptions of measurement of S in the literature [5, 6]. Last reference [6] gives description of an ac method that allows measurement of small samples and in a high pressure cell.

TEP can be measured by a 2-probe or 4-probe method. This is described in Figure 10.4. The basic difference between the two methods is the measurement of ΔT. In the 2-probe method the measured ΔT includes the ΔT created by the thermal boundary resistance of the sample and the end plates. In case of 4-probe method this problem does not arise, But the four probe method needs a somewhat larger sample. The use of 2-probe method thus would need that the thermal resistance of the two end contacts be much less than that of the sample. One should also keep in mind that the thermal boundary resistance may increase steeply (often as T^4) at low temperatures. So a low thermal resistance at room temperature or its vicinity is no guarantee for a low thermal resistance at low T. Often the bonding material may actually break and thus the thermal resistance can increase. This discussion is just to make the experimenter aware that use of 2-probe method needs caution.

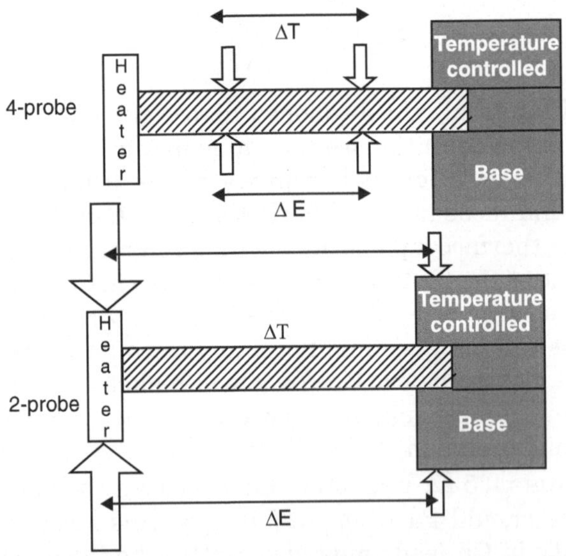

Figure 10.4: Schematic of 2-probe and 4-probe TEP measurements showing the position of thermal and electrical probes

In the following we describe a simple set-up for measurement of TEP [7]. The schematic diagram of the set-up is shown in Figure 10.5. This is a two-probe measurement. The base has a temperature controlled high conductivity copper plate. The temperature is controlled by an electronic controller. The extent of temperature control is mainly determined by the precision of the experiment needed. For instance if the relative precision expressed as a fraction is x, then typically one should measure the differential temperature ΔT with a precision much better than $x\Delta T$. This will thus demand that the temperature of the base is controlled to within $x\Delta T$. Typical good quality temperature controllers can control the base plate to within 0.01 K for temperatures less than 40 K, within 0.05 K for temperatures less than 100 K and within 0.2 K up to 400 K. This is much better than $x\Delta T$ in most cases.

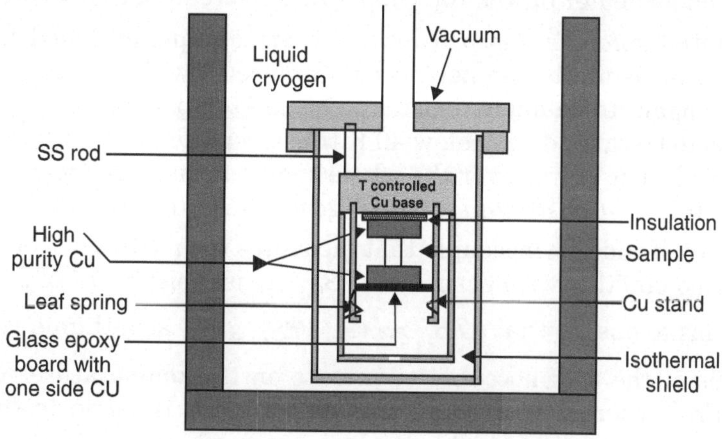

Figure 10.5: Schematic of a 2-probe measurements showing the details of constructions [7]

The base plate is attached to a copper plate that carries a copper shield that provides an isothermal environment around the sample. The copper plate carrying the shield, in turn, is attached to the flange of the cryostat using a thin stainless steel rod. All the electrical leads are thermally anchored to the shield. The thin stainless steel rod is needed to create a temperature differential between the base plate and the flange which is maintained at the cryogenic fluid bath temperature (T_B). (T_B = 77 K, when dipped in liquid nitrogen and T_B = 4 K when dipped in liquid He). For measurements from 80 K and up, liquid nitrogen can be used, and for measurements down to 5 K liquid He. In this context we point out the design change that is needed for measurements in closed cycle cryo-coolers that are very popular now-a-days. In this case, one may connect the sample holder plate directly to the cold finger of the cryo-cooler whose temperature is maintained constant and which can be changed independently. However, it is advisable to provide an isothermal shield around the sample.

The sample can be a pellet typically of diameter 10 mm, thickness 3-5 mm. It is held tightly between two polished copper discs (which can be electrochemically polished to maintain the shine) on the base plate using a glass epoxy plate with brass screws. The glass epoxy plate is an insulator and can bend to provide a pressure on the plate when the screw is tightened. It is like a "wide area" spring.

Pellet is not the only form one can use. One can use even small rectangular samples here provided one ensures good thermal contact of the two ends of the sample to the copper plates. The bottom copper disc is electrically insulated using a very thin cigarette paper attached with low temperature varnish like Ge-varnish. This is to ensure that the Cu plate is electrically insulated from the base plate but it is thermally anchored to it. It is better to float the sample electrically to measure the small thermo-voltage. An alternate solution for electrical insulation and good thermal contact is to have a thin plate of mica (for operation down to 80 K) or Sapphire (for operation down to 5 K) and hold it with *N*-grease. A small carbon resistor is used as heater on the top copper disc to create a temperature gradient.

A differential *gold (Fe doped)-Chromel* thermocouple attached to the copper discs is used to measure the temperature difference between the hot and cold ends. This choice is again to be made depending on the temperature range used. If the experiment is to be carried out below 40 K this is a preferred choice because it has a good thermoelectric power over the whole range. If the experiment is to be done only above 100 K *Cu-Constantan* is also a good choice. Typically we can make the choice this way. If the thermocouple that measures the ΔT has a thermopower S_{th}, then the thermo-emf that we measure is $\Delta T S_{th}$. ΔT is typically 5% of T as mentioned before. One has to ensure that $\Delta T S_{th} >> v_D$ ($\Delta T S_{th} \geq \frac{v_D}{x}$ as a thumb rule). The two chromel leads of the thermocouple terminate on the temperature controlled Cu base plate. From there a twisted pair of copper leads is taken to the top of the cryostat. From the top only shielded cables are used outside the cryostat to the voltmeter. Care is taken to ensure that the sample voltage leads and the thermocouple voltage leads are soldered using only low-thermal solder to minimize spurious voltages. Calibrated thermometers, attached to the base plate are used for measuring the sample temperature. The temperature difference and the thermoelectric voltage developed across the sample are measured using nano-voltmeters with GPIB interface.

After the stabilization of the base temperature, a small power $\dot{Q}$ is applied to the bottom copper disc (the one which is away from the base plate) using a heater attached to it. The power applied is determined by the thermal conductance of the arrangement and also the ΔT that we desire to create. Often it is best to apply the power at small intervals till one obtains a measurable ΔT. For computer controlled measurements one may use GPIB operated source. Else a small ramp being generated by a DA card and boosted by a current amplifier is a good solution. The temperature of the top copper disc increases steadily and consequently a small voltage will be developed across the hot and cold ends. V_{S-Cu} is the voltage developed across the Cu-Sample-Cu thermocouple for a corresponding voltage V_{Au-Ch} developed across the chromel-Au(Fe)-chromel couple. A plot of V_{S-Cu} vs. V_{Au-Ch} generally gives a straight line from whose slope one can find the quantity $S^* = V_{S-Cu}/V_{Au-Ch}$, which is used to find the TEP as described in the next paragraph.

If the heater power is sufficiently small, and the base temperature does not drift, $\Delta T/T$ is small (typically <5%) and the curve traces back in the reverse direction if the system is allowed to relax by switching off the heater power. In many situations an acceptable straight line (ΔT and ΔE curve) is not obtained or the heating or cooling curves show hysteresis. We found that such situations can come from the following reasons: (1) base temperature is not properly controlled; (2) ΔT is too small (typically $\Delta T/T < 1\%$ gives problem); (3) bad thermal contact between sample and copper discs and also between thermocouple junctions and copper discs; (4) widely different equilibration times for the sample and the thermocouple gives rise to hysteresis behavior if the heating rate is too fast compared to the larger of the time constants.

The slope of the heating (or cooling) curve gives the quantity S^*. This is the primary quantity that is measured by the experiment. To obtain $S_{S\text{-}Cu}$ we convert $V_{Au\text{-}Ch}$ to a corresponding ΔT at the measuring temperature $T + \Delta T/2$ where T is the base temperature. If $S_{Au\text{-}Ch}$ is the TEP of the Au-chromel thermocouple at temperature $T + \Delta T/2$, then $S_{S\text{-}Cu} = S^*.S_{Au\text{-}Ch}$. The absolute TEP of the sample S_S can be obtained from $S_{S\text{-}Cu}$ by adding S_{Cu} to it. S_{Cu} can be obtained from reported data [4].

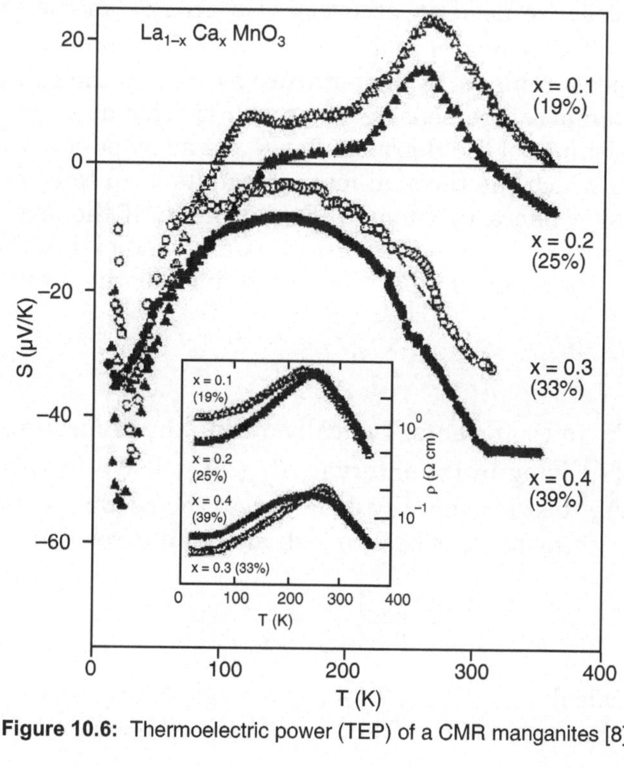

Figure 10.6: Thermoelectric power (TEP) of a CMR manganites [8]

A typical performance of the apparatus can be tested by measuring TEP of platinum in the range 5-300 K. For this purpose a length of a standard platinum wire (4 N pure) can be wound across a fused quartz disc, and this can be placed between the copper discs. The absolute value of TEP data obtained by the apparatus described above can be correct to better than 0.5 μV/K. The reproducibility of the measurement is also of the same order. We show in Figure 10.6 an example of the measurement of TEP made in colossal magneto-resistive (CMR) rare-earth manganites like $La_{1-x}Ca_xMnO_3$ ($x \approx 0.2-0.4$), which shows a large variation of TEP with temperature [8].

4. THERMAL MEASUREMENT (SPECIFIC HEAT AND THERMAL CONDUCTIVITY)

Measurement of thermal properties like specific heat (C_p), thermal conductivity (K) and diffusivity ($D \equiv K/\rho C_p$) are among the difficult measurements at low temperatures because both depend on the calibration of the thermometer as well as the fit error of the interpolation relation. The interpolation allows one to convert resistance (in case of resistance thermometer) or the voltage (as in the case of the diode) to temperature from the calibration curve which is obtained at discrete points. We can explain this in the following way. In the thermal measurements we measure the temperature increment ΔT after application of heat. Let us assume that $\Delta T/T$ = 0.05. In case we need an accuracy of even ±2% in the measurement we need to measure ΔT with an accuracy of ±0.1% of T. This immediately puts a large demand on the measurements of temperature as well as the calibration. Suppose it is a resistive temperature sensor. When the thermometer was calibrated we measured the resistance of the thermometer R_{th} at a few points within the interval of temperature in which the thermal measurements need to be done. In this case we fit an expression like a polynomial. For example, if the sensor is a resistive sensor the temperature T can be expressed as a polynomial of the resistance R_{th} over the temperature range of measurement defined as the interval $\{T_{\text{max}}, T_{\text{min}}\}$:

$$T = \sum_{n=1}^{N} \alpha_n R_{th}^n \tag{10.4.1}$$

Here the α_n are the fit coefficients. Typically during the calibration we have a set of $\{R_i\}$ for a set of $\{T_i\}$ lying in the interval $\{T_{\text{max}}, T_{\text{min}}\}$. We fit this data set to the above equation. A χ^2 test is generally done to test the goodness of fit. However, for the thermal measurements it is best to look at the fit error which we define as

$$\left|\frac{\Delta T_{\text{error}}}{T}\right| = \left|\frac{T_{\text{cal}} - T_{\text{obs}}}{T}\right| \tag{10.4.2}$$

where T_{cal} is the calculated value of T for a given value of R_{th} and T_{obs} is the observed temperature for that value of R_{th}. The fit will be good enough if the fit error is less than the desired accuracy that one needs for the measurement. It is a good practice to plot the fit error for the entire temperature range $\{T_{max}, T_{min}\}$ and check that the fit error is random (no systematic deviation), and over the whole range it is less than the desired accuracy. In the example we are using, in the entire range $\{T_{max}, T_{min}\} |\Delta T_{error}/T|$ should be $\leq 0.1\%$. The above example immediately shows that a good quality of the calibration of the sensor is the first starting point for thermal measurements. Any unattended error in this area can give wrong or noisy thermal data.

There are three other sources of error in most thermal measurements and an educated approach is needed to reduce these errors. These three sources are: (1) the thermal boundary resistance which often controls the thermal response/ relaxation time, (2) the mismatch of thermal response time of the sample and the response time of the thermometer attached to it (which in turn may depend on item (1)) and (3) spurious heat leak or input through measurement leads or radiation. There are various ways to determine the contribution of these sources.

4.1. The Time Constants

Every thermal measurement (like an electrical or a mechanical measurement) has time scales associated with it. The relative size of the thermal response times will decide the method that is used in the thermal measurements. The thermal response time will depend on whether we consider the body as *lumped* element or *distributed* element. This is explained in Figure 10.7.

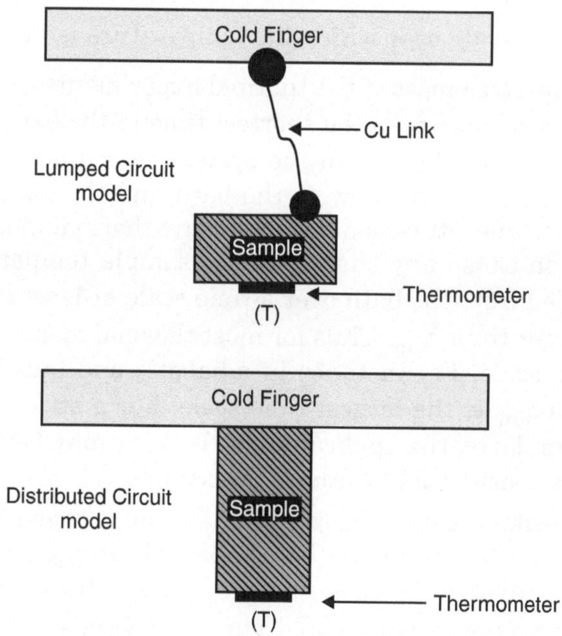

Figure 10.7: Two thermal models that are often used

Suppose we have a block of mass of heat capacity C connected to the cold finger by a set of Cu wires. This is often used when a rigid link cannot be used. Let the thermal resistance of the link be R. Then the response time of the Cu block to a heating or cooling pulse can be described by the lumped circuit model and often a single time constant is given by $\tau \equiv RC$. We also have situations where a sample of length L to be cooled is directly attached to the cold finger. In this case the thermometer attached to the sample will measure the temperature response that will follow heat diffusion equation. In this case the response will be in time scale $\tau \approx L^2/D$.

Let us consider four important time scales in the thermal measurements, namely:

(1) τ_h—the experimental scale over which the heat is applied to the sample. This is determined by the longer of the thermal time constants (Specific heat of the heater multiplied by the thermal resistance of the heater to the sample) and the scale of externally determined time programme that applies power to the heater.

(2) τ_s—the internal equilibration time of the sample including the thermometer. This is a function of sample size and the thermal diffusivity of the sample.

(3) τ_{bath}—the equilibrium time for the sample to relax to the bath. This is determined by the heat capacity of the sample and also the thermal resistance of the connecting link.

(4) τ_{exp}—the time scale over which the temperature is measured.

It is possible to analyze most of the thermal experiments in terms of the above time constants. In most cases τ_h is the shortest time in the experiment. In general if τ_{bath} is the shortest time element in the system we cannot create any thermal excitation because the heat will flow to the bath and the sample being strongly coupled to the bath temperature one cannot do any thermal measurements. So for thermal excitation to cause any change in the sample temperature, the sample needs to be decoupled from the bath over a time scale at least comparable to τ_{exp}, more preferably larger than τ_{exp}. Thus for most thermal measurements τ_{bath} may be the largest time scale. For instance in adiabatic and quasi-adiabatic specific heat measurements τ_{bath} is the largest time scale. For a strict adiabatic situation τ_{bath} tends to ∞. Similarly, the application of the heat may be done to the system before heat diffuses considerably through the sample.

In dynamic experiments one controls the heat input time scale. In this case τ_h need not be too short; but it should be comparable to τ_{bath}. In a.c. calorimetry, where a heater applies heat to a sample with a frequency $\omega_h(= 1/\tau_h)$, the temperature of the whole sample would vary following a frequency ω_h provided

$\tau_s^{-1} >> \tau_{exp}^{-1} > \tau_{bath}^{-1} > \omega_h$. In time varying thermal measurements, if the sample equilibrates within the time scale of the experiment, $\tau_s << \tau_{exp}$. Then we can use the sample as "lumped" circuit element (as in electrical circuits). If $\tau_s \sim \tau_{exp}$ the sample cannot be treated as a "lumped" thermal circuit and one needs to consider the thermal diffusion within the sample and the complete sample is not at the same temperature. This is the condition to measure the diffusivity D. Thermal measurements can be made very innovative and one may make them suit one's need if one has a thorough knowledge of all the time constants involved in the measurements.

In the following we present a simple example of how to use a lumped circuit model to understand heat capacity measurements. A simple *calorimeter* is shown in Figure 10.8. This calorimeter has been designed for use in the temperature range 0.5 K to 10 K [9]. However, with simple modifications it can be used even up to 300 K. The calorimeter consists of a sapphire (single crystal) plate of dimension 12 mm × 12 mm × 0.3 mm. The sapphire calorimeter is suspended from the thermal anchor plate (see Figure 10.8) by nylon threads or Teflon strands. The thermal anchor plate kept at base temperature T_B is the thermal ground for all leads (heater and thermometer) going to the calorimeter. One side of the sapphire plate has an evaporated Cr heater of resistance ~2 kΩ. The same side also has a carbon thermometer. The thermometer is an Allen Bradley resistor (10 Ω) that was ground down so that its room temperature resistance is around 1–1.2 kΩ. The carbon thermometer is mounted on the sapphire calorimeter by little GE 7031 varnish (A better alternative is to evaporate Ge-Au alloy strip on sapphire for use as thermometer). The leads to the heater and thermometer are made from 75 μ superconducting Nb wire. This gives leads of low thermal conductance and extremely high electrical conductance. The use of low thermal conductance Nb leads allows us to tune the thermal link (described below) to proper K value by a separate copper wire. The ends of the Nb leads have small pieces (1-2 mm) of Cu-Ni capillary crimped on to them which serve as soldering lugs. However the operation of the calorimeter is restricted to below T_c of Nb (≈9.6 K). The thermometer is calibrated with all the leads that are actually used, to avoid any uncertainties.

The thermal link of the calorimeter is made of Cu wire and is directly connected to the sample with a small piece of Cu foil. Having a proper length of the Cu wire is an important part of the design because this determines the constant for relaxation to the bath. For measurements down to 0.5 K experiments were carried out in a simple home made He^3 cryostat. The calorimeter described here depends on the well-known principle of relaxation calorimetry and adiabatic and differential scanning calorimtery [10-13].

The thermal model is shown in Figure 10.9. This is similar to the model used in Figure 9.12 for the temperature controller. Here C_A is the total heat capacity of the calorimeter plate including that of the thermometer and the heater. We assumed

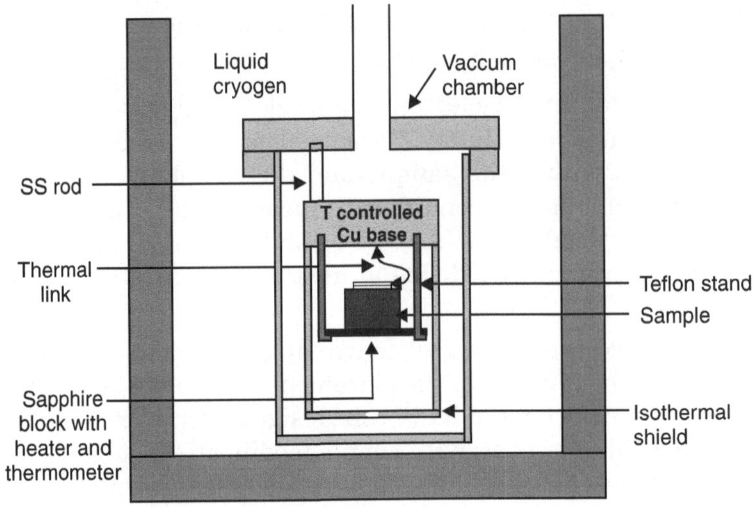

Figure 10.8: A simple calorimeter to measure heat capacity

that the heater and the thermometer mounted on the calorimeter have good thermal contact with the calorimeter plate and thus the equilibration time constant of them to the plate is very small. R_B is the boundary resistance between calorimeter and the sample. Aim is to make this small compared to the other thermal resistances like R_L or R_S described below. C_S is the sample heat capacity and R_S is the sample thermal resistance. R_L is the thermal resistance to the bath. This thermal resistance in the final analysis will control the relaxation time constant when a heat pulse is applied to the system. We can further simplify the above model as below by redefining $R_1 = R_B + R_S/2$, $R_2 = R_L + R_S/2$.

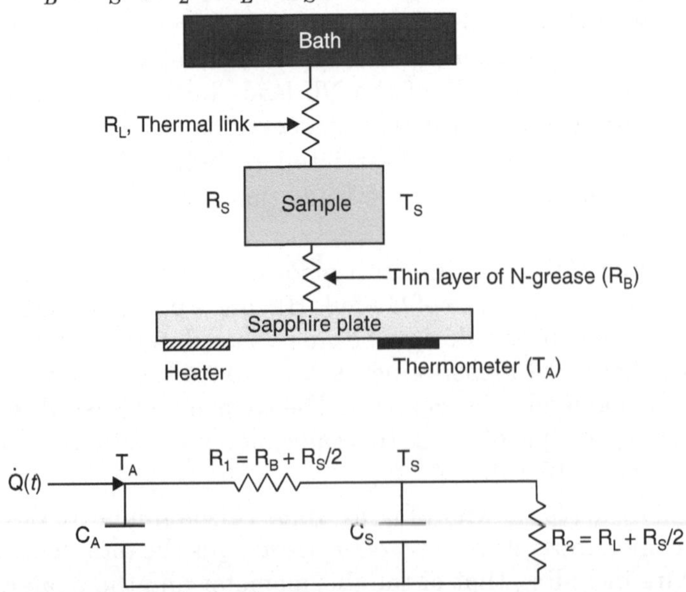

Figure 10.9: A thermal model of the calorimeter to shown in Figure 10.8

In the above model the temperature that we actually measure is T_A measured on the calorimeter, while T_S is the actual temperature of the sample. (All temperatures are measured with respect to the thermal ground that is the base temperature of the cryostat.) Suppose the heat current given to the calorimeter is $\dot{Q}(t)$. We would like to investigate the response of the calorimeter carrying the sample to such a heat current.

We write the heat balance equations that are like the loop equations in Kirchoff's law.

$$\dot{Q}(t) = C_A \frac{dT_A}{dt} + \frac{T_A - T_S}{R_1} \tag{10.4.1}$$

$$\frac{T_A - T_S}{R_1} = C_S \frac{dT_S}{dt} + \frac{T_S}{R_2} \tag{10.4.2}$$

By using the above equations we can establish a very important relation. When a heat current of finite duration is applied to the calorimeter, so that

$$\int_0^\infty \dot{Q}(t)dt = \Delta\text{H} = finite \tag{10.4.3}$$

The temperature of the sample T_S as a function of time varies in such a manner that we also have:

$$\Delta\text{H} = \int_0^\infty \dot{Q}(t)dt = \left(\frac{1}{R_2}\right)\int_0^\infty T_S(t)dt = C_S \int_0^\infty T_S\left(\frac{t}{\tau}\right)d\left(\frac{t}{\tau}\right) \tag{10.4.4}$$

The above result indicates that the area under the time-temperature curve does not depend on actual time or mode by which energy is supplied to the sample. This is because ultimately all the heat energy passes to the bath through the sample. The area under the time-temperature curve will always remain the same. (Here $\tau \equiv C_S R_2$, is the time constant for relaxation of the sample to the bath). Equation (10.4.4) gives us a direct way to find the heat capacity from the time-temperature curve which, being digitized, can be easily integrated. The time constant τ can be determined easily from the experiment. In reality we actually measure T_A. By integration of equation (10.4.2) one can relate the area under the (T_A vs. t/τ) curve to the heat capacity C_S:

$$\Delta H = C_S(1+\lambda)^{-1}\int_0^\infty T_A\left(\frac{t}{\tau}\right)d\left(\frac{t}{\tau}\right) \tag{10.4.5}$$

Here $\lambda \equiv R_1/R_2$ and generally $\lambda << 1$.

The heat capacity is generally determined by applying a square heat pulse (power P_0) over a duration t_0. The time temperature curve ($T_A - t$) is recorded by a computer and an analysis of the curve gives the heat capacity. The resulting curve shows a peak and then decays following an exponential law to the bath temperature at $t >> \tau$. The coupled differential equations (10.4.1) and (10.4.2) can be solved by using Laplace transform and one can obtain an expression for T_S and T_A as a function of time following the heat pulse. The heat capacity can be determined by finding the peak temperature or from the decay curve following the equations given below which is valid for $t > t_0$ (i.e., after the end of the heat pulse given to the sample):

$$t_A(x) \cong \left(\frac{e^{x_0}-1}{x_0}\right)\left[(1+r)-\lambda_s\tau_1\right]e^{-x} - \left(\frac{e^{\lambda_r x_0}-1}{\lambda_r x_0}\right)\left[(1+r)-\lambda_r\lambda_s\tau_1\right]e^{-\lambda_r x} \quad (10.4.6)$$

where the temperature T_A (which is the temperature actually measured by the thermometer as distinct from the sample temperature) and the time are scaled to get the dimensionless parameters t_A and x and the response is described by two exponential terms. For longer times ($\lambda_r x >> 1$), equation (10.4.6) reduces to a single exponential controlled by a single parameter λ_s. The terms in equation (10.4.6) are defined as:

$$c \equiv \frac{C_A}{C_S}, \quad r \equiv \frac{R_1}{R_2}, \quad t_A \equiv T/T_0, T_0 \equiv \frac{P_0 t_0}{C_S}, \quad x \equiv \lambda_S t, x_0 \equiv \lambda_S t_0,$$

$$\tau_1 \equiv C_S R_1, \; \tau_2 \equiv C_S R_2, \; \beta' \equiv c(1+r)$$

with the two time constants defined as:

$$\lambda_s \equiv \gamma - \sqrt{\gamma^2-\delta^2}, \qquad \lambda_r \equiv \frac{\gamma+\sqrt{\gamma^2-\delta^2}}{\gamma-\sqrt{\gamma^2-\delta^2}} \quad (10.4.7)$$

with $\delta^2 \equiv \dfrac{1}{c\tau_1\tau_2}$ and $\dfrac{\delta^2}{\gamma^2} \equiv \dfrac{4cr}{(1+\beta')}$. Generally c as well as r are << 1 so that

$\lambda_S \approx 1/2\gamma\left(\dfrac{\delta^2}{\gamma^2}\right) = \tau_2^{-1}$. In the ideal case when $R_2 \approx R_L$, the time constant is the expected value for the relaxation of a thermal mass of heat capacity C_S through a thermal link R_L. In this case the difference between T_A and T_S is so small and we get the simple relation for ($t >> t_0$):

$$t_A \approx e^{-x}, \qquad \text{or} \qquad T_A \approx (\Delta H/C_S)e^{-\frac{t}{\tau_2}} \quad (10.4.8)$$

with $\Delta H \equiv P_0 t_0$, the total heat energy supplied to the system. Equation (10.4.8) is most widely used in relaxation calorimetry. However, it should be pointed out that it is an approximation to the exact relation given by equation (10.4.6) and the conditions for the approximation, $r << 1$ and $c << 1$ and $R_B << R_L$ need be strictly satisfied during the experiment.

Measurement of heat capacity under different conditions and techniques needs a thorough understanding of the heat relaxation process and this needs to be properly incorporated in the measurements.

In recent years an exciting technique called 3ω *technique* [14, 15] has been proposed as a modification of conventional ac calorimetry [16]. In this technique one actually measures the temperature oscillations of a heater kept in contact with a medium and obtains information on the thermal diffusion length. From this one can find the thermal diffusivity. The basic concept is described in Figure 10.10.

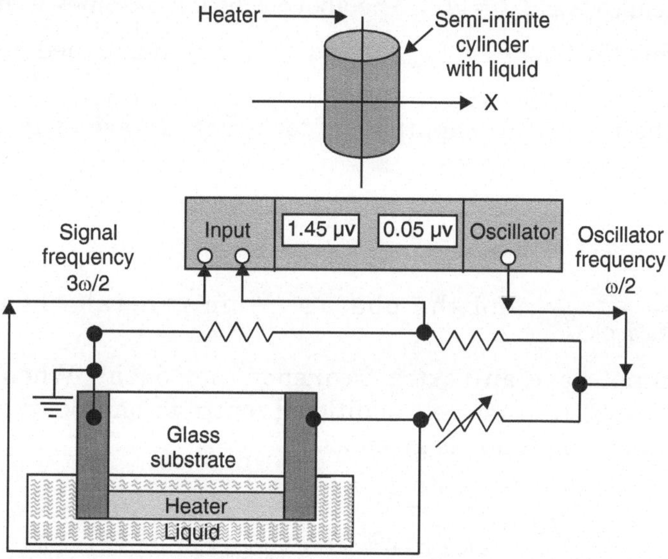

Figure 10.10: A Schematic of the 3ω technique (adapted from [14])

A substrate (like a glass) has an evaporated thin heater which is immersed in the liquid or the medium whose diffusivity has be measured. The heater is driven by a current at frequency ω/2. The current through the heater will heat it up at frequency ω by Joule heating. Due to a finite temperature coefficient of resistivity of the heater the voltage across the heater will have a 3 ω/2 term in addition to the regular ω/2 component. The ω/2 component is largely suppressed by a bridge arrangement and the 3 ω/2 component is detected by a lock-in amplifier. One can obtain information on the diffusivity D from this component as shown below [14, 15].

The temperature of oscillation (sinusoidal) around a wire dipped in a liquid (heat capacity C_p, thermal conductivity κ, density ρ) will follow the one-dimensional heat flow equation [17]:

$$-j\omega T(x,\omega) = D\frac{\partial^2 T}{\partial x^2} \tag{10.4.9}$$

where $D \equiv \dfrac{\kappa}{\rho C_p}$ is the diffusivity of the surrounding medium. The solution of the above equation is determined by the complex wave vector:

$$k = \left(\frac{\omega}{2D}\right)^{1/2}(1-j) \tag{10.4.10}$$

The complex wave vector introduces a length scale to the problem, namely the thermal wavelength $\lambda_{\text{thermal}} = |k|^{-1}$. The above solution assumes a thin heater that is valid if the film thickness << λ_{thermal} and the surrounding medium is considered infinite when the dimension of the container is >> λ_{thermal}. The temperature oscillation of the wire with amplitude ΔT for a heat flux density (Watt.m^{-2}) $\dot{Q}$ is given as:

$$\Delta T(0,\omega) = \Delta T_0 e^{j\varphi} \tag{10.4.11}$$

where $\Delta T_0 = \dfrac{\dot{Q}}{(\omega c_p \rho \kappa)^{1/2}}$ and the phase $\varphi = \dfrac{\pi}{4}$ in general when the thermal parameters do not have any extra frequency component. When the heater is evaporated on a substrate one gets additional contribution from the substrate and then the oscillation amplitude is given as:

$$\Delta T_0 = \frac{\dot{Q}}{(\omega c_p \rho \kappa)^{1/2} + (\omega c_{\text{sub}} \rho_{\text{sub}} \kappa_{\text{sub}})^{1/2}} \tag{10.4.12}$$

The subscript "sub" signifies the thermal parameters for the substrate. The experimental arrangement shown in Figure 10.10 is used to find the magnitude of temperature oscillation. The temperature oscillation at frequency ω creates an oscillation of the resistance at frequency ω, given as:

$$\Delta R(\omega) = \Delta T_0 \beta R_0 e^{j(\omega t - \varphi)} \tag{10.4.13}$$

The coupling of the heater resistance to the temperature oscillation is through the temperature coefficient of resistivity $\beta \equiv \dfrac{1}{R}\dfrac{dR}{dT}$. The heater material should have a large β. The current through the heater is given as $I = I_0 e^{j\omega t/2}$. Taking the real part of the voltage oscillation due to the thermal oscillation that is at frequency $3\,\omega/2$, we have:

$$\Delta V(3\,\omega/2) = \frac{1}{2}\Delta T_0 \beta I_0 R_0 \cos(3\,\omega t/2 - \varphi) \tag{10.4.14}$$

From the amplitude of the detected voltage using equations (10.4.12) and (10.4.13) one can find out the value of the parameter $(C_p\kappa)^{1/2}$. An independent knowledge of either κ or C_p will give the other quantity.

An important modification of the technique has been suggested to find the thermal conductivity κ of a thin film by using a narrow heater [15].

5. MAGNETIC MEASUREMENTS

Magnetic measurements are one of the most fundamental measurements that fall in the category of low temperature measurements of properties of materials. Good magnetic measurements are not only important for the field of magnetism. Often a magnetic system can be a physically realizable model of a physical theory. A good magnetic measurement thus can be used to test a basic theory. Magnetic measurements are also important for characterization of technologically important magnetic materials that can be produced in bulk like ferrites, ferroalloys or superconductors, or in thin film form as those used in memory devices. The measurement of magnetic properties is a rather specialized topic and a detailed discussion is beyond the scope of this book.

The basic measurements of magnetization M are based on two fundamental methods, namely the force method and the induction method. In the force method the most popular one is the *Faraday Balance*. In this method one places the sample (magnetized by a d.c. field B_a to a total magnetization μ_{tot}) in an inhomogeneous field (field with a gradient) and then measures the force on it by a precision balance. In this case, when there is a gradient of the field in the z-direction, the force on the magnetic material is in the z-direction. It is given by:

$$\vec{F} = \mu_0 \mu_{tot}(B_a)\frac{\partial B_a}{\partial z}\hat{z} \qquad (10.5.1)$$

The anisotropy in magnetization when the sample is placed in a magnetic field gradient gives rise to a torque on the specimen which can also be measured.

A very interesting development in the force method is the *alternating gradient magnetometer*. This is particularly suitable for small samples. In this method the force is measured by a cantilever. A resonance detection scheme is used that enhances the sensitivity to the range of 10^{-8} emu. In this method the sample is magnetized by a d.c. field as in the Faraday Balance. However, the gradient field that produces the force is made to alternate at a frequency close to the resonance frequency of the cantilever that carries the sample. This being a forced oscillation the amplitude of oscillation is proportional to the force and hence to the magnetization of the sample. Thus the amplitude of oscillation is a direct measure of the sample magnetization. The amplitude of oscillation can be detected by a number of methods including optical and piezoelectric methods. The resonance enhances the amplitude of oscillation of the cantilever for a given force and thus enhances the sensitivity. The advantage of this method is that the cantilever can be miniaturized by the techniques of Quartz or Silicon micromachining and thus one can use micron-sized samples. Advantage of quartz resonator is its high quality factor. A sensitivity of nearly 10^{-10} emu has been achieved with micro-cantilever based magnetometers and an optical interference detection scheme.

The second method is based on induction as in the vibrating sample magnetometer (VSM) or a SQUID. Most of the practitioners of measurement of magnetic properties now a days use commercial versions of VSM and SQUID. There are however simpler descriptions of these apparatus that still follow the "home-built" tradition and some of them can achieve a sensitivity better than commercial instruments. We direct the readers (even those who use or plan to use commercial set-ups) to these articles mainly to demystify the black boxes. Below we give simplified descriptions of some of these set-ups that are mainly used for low temperature measurements.

5.1. Rudiments of VSM

The vibrating sample magnetometer (VSM) is a simple yet effective technique for characterizing properties of magnetic materials. With proper calibration it gives a direct measure of the Magnetization (M). It was first made by *S.* Foner [18]. The principle of the VSM is simple and is a direct application of Faraday's Laws of induction. A magnetic sample is placed in a magnetic field H so that it makes the magnetic sample a magnetic dipole of moment $\boldsymbol{\mu}$. The sample is kept at the end of a long rod and then driven by a mechanical vibrator at frequency ω. The vibration frequency is typically in low audio frequency range. The oscillatory motion of the magnetized sample moment of μ will induce a voltage in the detection coils. The induced voltage is proportional to the magnetization M of the sample under study.

Design and positioning of the detection coils may be the most important aspect of VSM design. We refer to some of the published references for the details [19, 20]. In the following we give a simplified treatment that allows one to appreciate the issues involved and how a simple design can be made [21].

The sample which is magnetized is approximated as a dipole of moment $\boldsymbol{\mu}$. This is definitely a first level of approximation and in more demanding use of VSM one should have the multipole expansion. If we have an oscillating dipole of moment $\boldsymbol{\mu}$, this creates an oscillating magnetic field and the question we ask is at what point should a coil be placed to create the maximum induced emf. The field of the dipole is given as:

$$\vec{B}(r,0) = \left(\frac{\mu_0}{4\pi}\right)\left(\frac{\mu}{r^3}\right)(2\cos\theta\,\hat{r} + \sin\theta\,\hat{\theta}) \qquad (10.5.2)$$

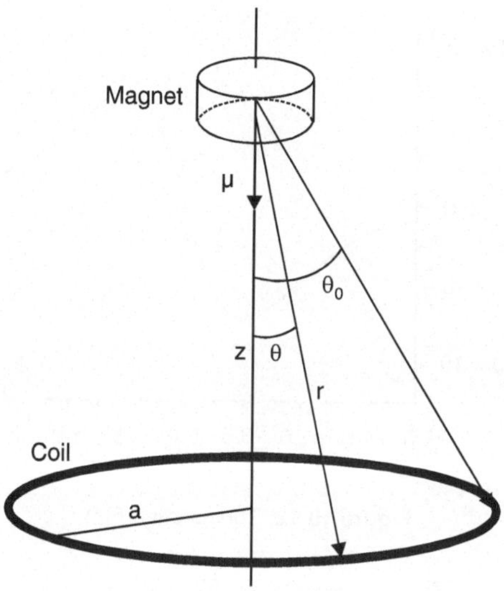

Figure 10.11: The coordinates used and the magnetic moment

The integrated flux Φ at the coil (with N turns and radius a) placed at a distance r from the dipole and kept with its axis parallel to the axis of the dipole moment is given as:

$$\Phi = \int \vec{B}.d\vec{A} = \mu_0\left(\frac{\mu N}{2r}\right)\sin^2\theta_0 = \mu_0\left(\frac{\mu N}{2}\right)\frac{a^2}{(a^2+z^2)^{3/2}} \qquad (10.5.3)$$

where the angle θ_0 is defined in the diagram and $\sin\theta_0 = (a/r)$ and $r = (a^2 + z^2)^{1/2}$, z being the vertical distance of the dipole from the coil. If the magnetic dipole moves at a velocity $v(\equiv dz/dt)$ along the axis the induced emf at the coil, expressed in a reduced scale $\xi \equiv z/a$, is given as

$$E_{\text{induced}} = -\partial\Phi/\partial t = -\left(\frac{\partial\Phi}{\partial z}\right)v = \frac{3}{2}\mu_0\left(\frac{\mu N}{a^2}\right)\frac{v\xi}{(1+\xi^2)^{5/2}} = \frac{3}{2}\mu_0\left(\frac{\mu N}{a^2}\right)vF(\xi) \qquad (10.5.4)$$

It is interesting to note that for a given ξ the induced voltage is $\propto 1/a^2$. This is counter-intuitive as one would expect a bigger coil to produce a larger emf as the flux enclosed will be more!

For a given a, the coil radius, E_{induced} increases with ξ and is $\propto \xi$ for $\xi << 1$. For $\xi >> 1$, E_{induced} decreases as ξ increases. The optimum value of E_{induced} is obtained at $|\xi| = 1/2$ where the function $F(\xi)$ reaches a maximum ≈ 0.28. The maximum induced emf is thus given as:

$$E_m \approx 0.43\mu_0\left(\frac{\mu N}{a^2}\right)v \qquad (10.5.5)$$

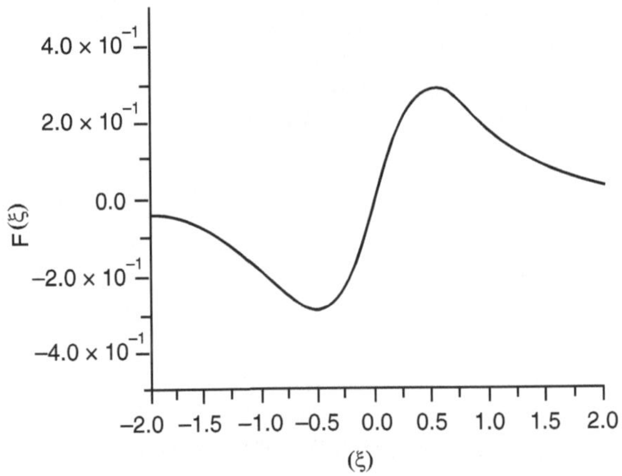

Figure 10.12: The function $F(\xi)$

A plot of the function $F(\xi)$ is given in Figure 10.12. One can see that for this geometry used the induced emf is maximum at $|\xi| = ½$. There are two symmetrical positions for positive and negative values of ξ where the induced emf has the maximum magnitude. One can increase the induced emf by a factor of 2 by placing two coils one at $\xi = 1/2$ and other at $\xi = -1/2$ and then add up the induced emf of the two by placing them in series. When the vibrating sample executes a simple harmonic motion at a frequency f (as is the case more often), we get the displacement $\Delta z = A_0 \sin(2\pi ft)$, A_0 being the amplitude. So $v = (2\pi f)A_0 \cos(2\pi ft)$. We will then get a sinusoidal induced voltage (provided $A_0 << z = a/2$) that is given as:

$$E_m \approx 5.4\mu_0 f\left(\frac{\mu N}{a^2}\right)A_0 \cos(2\pi ft) \quad (10.5.6)$$

It is interesting to estimate the minimum detectable μ for such a system of two coils placed at $\xi = \pm 1/2$. Let us make a reasonable assumption that $N = 1000$, $a = 15$ mm and $A_0 = 1.5$ mm and $f = 100$ Hz. For the electronics that uses a lock-in amplifier, we can put a resolution of voltage of 100 nV. The minimum detectable magnetic moment $\mu \approx 2.5 \times 10^{-5}$ emu.

The calibration of the VSM is generally done by using standard samples whose moments are known. Method of calibration of VSM is a serious task if one needs quantitative data. This has been discussed elsewhere [20].

5.2. A.C. Susceptibility

In addition to magnetization measurement another magnetic quantity that is often measured is the ac susceptibility, χ, which is defined as the derivative of magnetization μ with respect to H and is often referred to as the dynamic susceptibility.

$$\chi(H) = \frac{\partial \mu}{\partial H} \tag{10.5.7}$$

χ is measured in an applied d.c. field. In fact μ can be reconstructed (to the uncertainty of the residual magnetization) by a numerical integration of the $\chi(H)$ curve. However, in most measurements of ac susceptibility the applied field is zero and one gets the zero field susceptibility.

Measurement of AC susceptibility (low frequency) by using a mutual inductance bridge is a popular, and, may even be, the cheapest method to characterize magnetic materials. This method, unlike techniques like the Faraday balancc or VSM, gives the dynamic susceptibility $\chi \equiv \partial M/\partial H$. Though this method has less sensitivity than the other, it has the advantage of giving the zero-field susceptibility and also involves simpler instrumentation. Most of the AC mutual inductance bridges reported [22, 23] are based on some modification of the basic design of Hartshorn [24]. The various modifications introduced led to certain advantages in measurement or an increase in the sensitivity. What we describe below is a realization of the basic ideas of Hartshorn Bridge and by no means is it unique [25].

In Figure 10.13, the schematic of the bridge is shown [25]. The output of the internal oscillator of the lock-in is amplified by an audio power amplifier and is used to excite the bridge. The transformer T_2 provides the ground isolation. The current i_p, which passes through the primary of the mutual inductance M, also passes through a resistor R_2. (The coil design depends on the sample. One needs to have a balanced pair of secondaries. Typical number of turns can be around few thousands to even more than ten thousand. The self-inductance of primary and the individual secondaries are generally about few tens to few hundreds of milliHenries. The two secondaries can be balanced to even 1% or better thus reducing the coil contribution to a small value.) A precise fraction of the voltage is fed into the operational amplifier A_1 with the help of a precision voltage divider. For manual balancing this is the part that is used for balancing. If automation is needed this is the part where one can incorporate multiplying DAC for automating balancing. The primary of a unity gain transformer T_1 is put in the feedback loop of A_1. This ensures that the voltage V_s induced in the secondary of the transformer (which is connected in series with the secondary of the mutual inductance coil) is such that it has a phase difference of $3\pi/2$ with respect to the voltage across R_2 and hence i_p. This practice of having an inductive element in the feedback loop should be followed with caution because this some times make this circuit unstable towards oscillation.

The voltage induced in the mutual inductance coil is given by

$$V^* = (R + j\omega M)i_p \tag{10.5.8}$$

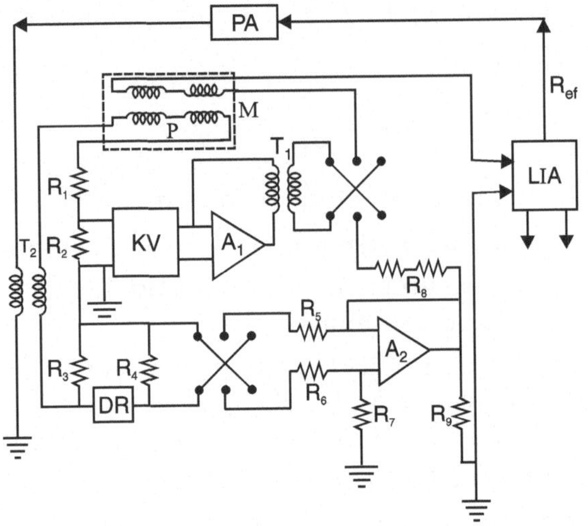

Figure 10.13: Schematic of a typical mutual inductance bridge (redrawn from ref. 25)

$(R + j\omega M)$ is the impedance of the coil at frequency $\omega = 2\pi f$. The resistive or dissipative part contains the resistance (d.c.) of the coil and a part $\omega M''$ where $M'' \propto \chi''$, the imaginary part of the susceptibility. One part of this voltage which is proportional to the mutual inductance of the coil (and hence the real part of the susceptibility (χ') of the material) will have a phase difference of $\pi/2$ with respect to i_p. The voltage divider transformer combination acts as a variable mutual inductance. The voltage $\omega M i_p$, which contains the information about χ' is balanced by V_s.

The other part of V^*, which is proportional to the imaginary part of the susceptibility (χ''), can be similarly balanced by a simple voltage divider circuit containing a decade resistance box (DR). The decade box varies the voltage across the resistance R_4 which is sensed by a unit gain differential amplifier A_2, which adds (or subtracts) the voltage from the secondary circuit. An alternate arrangement is to use another precision voltage divider and to balance off V^* by a voltage developed across resistor R_3. The output of the bridge is detected by a two-phase lock-in amplifier and the bridge is balanced. Balancing the bridge electronically using an op-amp feed back by tapping a part of the primary voltage is a common concept that is used in most of the a.c. susceptibility set-ups.

Mutual inductance bridge for a metallic object gives a contribution arising from the eddy current in the metal due to the ac field of the primary. This can be measured and provides a way to measure the resistivity of conducting solids by a contact-less method [26-29].

A very useful magnetic quantity that can be measured using the mutual inductance bridge is the non-linear susceptibilities $\chi^{(n)}$ which are higher order susceptibilities defined as:

$$\chi^{(n)} \equiv \frac{\partial^n \mu}{\partial H^n} \tag{10.5.9}$$

For materials like spin-glass the third harmonic diverges in the limit of zero applied field. The higher order susceptibilities show up as higher harmonics in the induced voltage of the secondary. In most modern lock-in amplifiers, that are DSP-based, a large number of higher harmonics can be detected. Caution should be exercised that the input exciting signal should be mostly free of higher harmonics. This can be tested by an empty secondary coil.

The arrangement for the cryostat is shown in Figure 10.14. The mutual inductance coil is kept at constant temperature by dipping it directly into a cryogenic bath (liquid nitrogen or liquid helium). Both the primary and secondary coils are wound on a fibre-glass cylinder which fits tightly around a double-walled glass tube. For smaller coils one can also use Hylam or even good quality teak wood. The double-walled tube provides the necessary thermal isolation for varying the temperature of the sample which is kept in the innermost tube. The sample holder is held from the top by a support rod of Hylam. The sample can be moved in and out of the coil by moving this rod through a O-ring coupling.

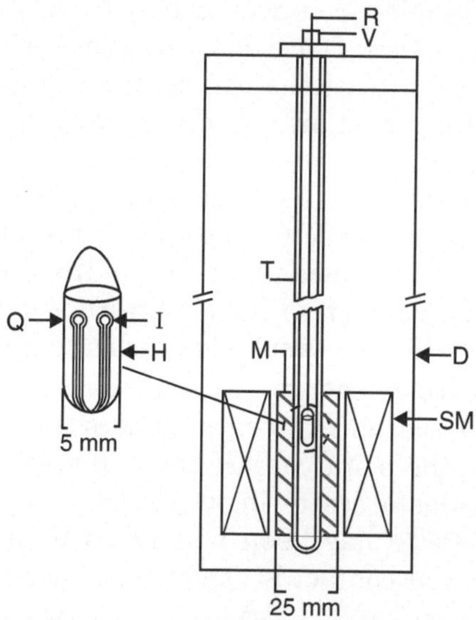

Figure 10.14: Schematic of a typical cryostat for mutual inductance bridge (ref. 25)

The details of the sample holder design depend on the type of the sample. In the double-walled tube, the space between the walls can be kept evacuated all the time or a small amount of exchange gas can be introduced in this space depending on the need. The sample space can have He gas at atmospheric pressure. This ensures that the sample is at uniform temperature and the thermometer (typically a Pt 100 or a Cernox both of which have a small dependence on the magnetic field) can be kept at a distance and outside the coil. The sample and thermometer then can be connected by a highly thermally conducting sapphire rod.

How low a value of susceptibility χ can be measured with an a.c. bridge? This depends on the magnitude of the induced voltage, the amplification and how that compares with v_D. Typically a value of $\chi \approx 10^{-3}$–10^{-4} emu can be obtained with most set-ups. With some care this can be pushed down by two orders. This of course is difficult to do but can be achieved provided special care is taken to shield spurious pick ups and reduce drifts. The induced voltage in the coil will depend on the number of turns in the coil, its geometry and the fill fraction, which is the fraction of volume of the coil being occupied by the sample.

Measurement of magnetic properties in materials is a knowledge intensive activity. It is best done when the experimenter has a clarity on basics of measurements. The reader is referred to some classic books like ref. [30] that should be made a reference for any magnetic measurements.

6. POINT CONTACT SPECTROSCOPY

Point contact spectroscopy [31-32] is an elegant technique for studying electronic transport properties across micro-contacts. In particular, this technique can provide direct information on interaction of electrons with other quasi-particles like phonons and magnons and in certain cases one can obtain a quantitative measure of the spectrally resolved coupling constant like the electron-phonon coupling constant. Briefly when the micro-contact between two metals is sufficiently small the $I-V$ characteristics of such metallic point contacts show markedly different behaviour from those of bulk or ohmic contacts. One may think of the micro-contact as a circular orifice of radius r. If the mean free path of the conduction electron, $l_{el} > r$, the electron can cross the contact without scattering and its motion is ballistic. When bias (a potential difference) is applied in such a contact the electron is injected into the junction region with an energy $\approx eV$ where V is the applied bias. In absence of such ballistic condition, scattering always reduces the energy to the scale of k_BT. The ballistic injection in the point contact thus allows for direct energy-resolved information about the inelastic scattering of electrons in the contact region. Point Contact Spectroscopy (*PCS*) has been widely used to study electron-phonon interaction in solids. A point contact is generally characterized by its size that in turn is measured by the Knudsen number $\kappa \equiv l_{el}/r$. Most point contacts that are used for PCS are ballistic in nature with $\kappa \geq 1$. In this regime the resistance is given by the *Sharvin resistance* that is defined as [31, 32]:

$$R_S = \frac{4\rho_0 l_e}{3\pi r_c^2} \tag{10.6.1}$$

In this case ρ_0 is the resistivity of the material and $A = \pi r_c^2$ is the area of the point contact. (In the opposite limit of large contact $\kappa \to 0$, one has the Maxwell regime. In this regime due to multiple scattering the electron completely loses its energy resulting in Joule heating of the junction. Assuming a circular orifice, Maxwell derived the expression, $R_M = \rho_0/2r$, for the junction resistance (ρ_0 is the resistivity of the material)).

The information about the PCS spectra is obtained from the double derivative d^2I/dV^2. The double derivative is finite due to inelastic interactions of the electrons with other elementary interactions like phonons. The strength of the inelastic interaction is measured by the spectral function $\Im(\varepsilon)$. This incorporates any inelastic interactions due to phonons, magnons or magnetic impurities. For a metallic junction with DOS (density of states) $N(0)$ at the Fermi level, the relation of d^2I/dV^2 with $\Im(\varepsilon = eV)$ is given as [31, 32]:

$$\frac{d^2I}{dV^2} = -\left(\frac{2\pi e}{\hbar}\right)\Omega_{eff} N(0)\Im(eV) \tag{10.6.2}$$

For electron-phonon interaction the *spectral function* $\Im(\varepsilon)$ is the *Eliashberg function* $\alpha^2F(\omega)$. Ω_{eff} is an effective volume within which the inelastic scattering takes place. For the ballistic regime, $\Omega_{eff} = 8r_c^3/3$. Figure 10.14 shows the Eliashberg function of a Au-Au point contact junction obtained from the PCS data. This Elisahberg function obtained from PCS differs from the actual Eliashberg function by a form factor η so that $\alpha^2F(\omega) \cong \alpha_{pc}^2 F(\omega)/\eta$. The exact value of η depends on the geometry of the contact and the Knudsen number. In the ballistic regime, $\eta \approx 1/4$. In addition, further complications arise from non-equilibrium phonons. These make additional contributions to the double derivative as a background. The merit of this method lies in the fact that it gives the Eliashberg function $\alpha^2F(\omega)$, which is the electron-phonon interaction function for a metallic system. This in turn can be used to calculate the el-ph coupling constant λ, which we define as:

$$\lambda = 2\int_0^{\omega_m} \frac{\alpha^2F(\omega)}{\omega}d\omega \tag{10.6.3}$$

ω_m is the maximum phonon frequency. PCS spectroscopy can be used also for complex systems and one can obtain the contribution of phonons to the resistivity.

Phonons are not the only thing that can be studied by PCS. There are quite a few other interesting physical problems in solid state physics that can be fruitfully investigated by the PCS. This method gives us a way to inject an electron inside a solid with the energy defined by the applied bias. The electron then can interact with other excitations and from the non-linear conductivity one can obtain useful information. One of the interesting applications of the PCS is in studies of superconductivity and such phenomena as Andreev reflection [33, 34]. Magnetism and interaction of conduction electrons with spins like Kondo effect is yet another phenomenon that can be studied by using this method.

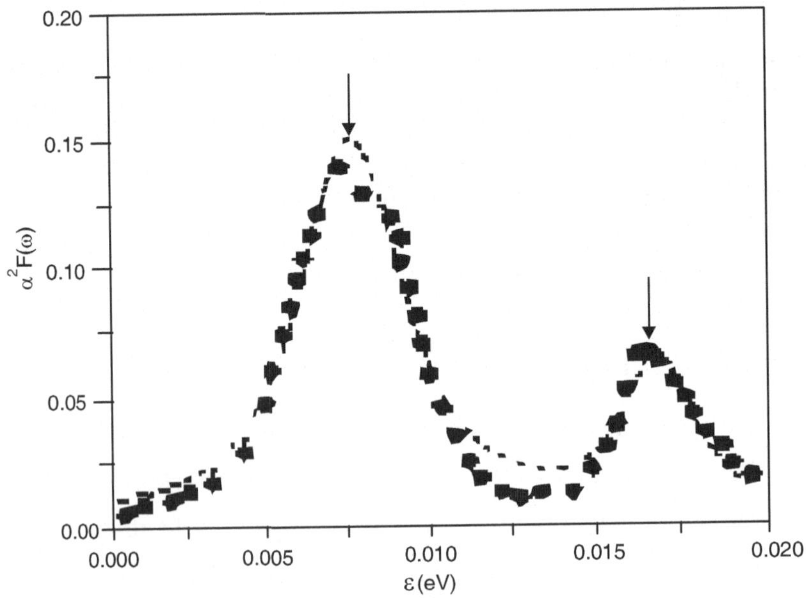

Figure 10.15: Phonon spectrum of Au obtained in Au-Au point contact [36]

6.1. Experimental Details of PCS

Here we present an overview of the mechanical and electronic instrumentation involved in making a point contact junction and measuring the junction conductance as a function of bias, with a change of temperature and magnetic field [35, 36]. We then provide the example of point contact tunneling data. The most common way of acquiring the PCS data is to directly measure the differential conductivity, $g = dI/dV$ as a function of V (g-V curve) using an AC modulation technique. Some of the experimentalists measure the differential resistance dV/dI. This can be easily converted to dI/dV and vice-versa. One can also use a d.c. technique and measure the I-V curve from which g can be obtained by numerical differentiation. When measurements are done with the bias within tens of mV, with an energy resolution of a fraction of a mV, and when the absolute change in dI/dV is only a few percent, AC modulation technique gives a higher signal to noise ratio compared to its DC counterpart. Experimentally, we apply a constant amplitude AC modulation voltage $\delta V \cos(\omega t)$ overriding a DC bias, V_{DC} across the junction. The voltage across the junction is $V = V_{DC} + \delta V \cos(\omega t)$ with $\delta V << V_{DC}$. The DC bias across the junction determines the energy of the electron crossing the barrier. A Taylor series expansion of the total current across the junction gives,

$$I(t) = I_{DC} + \frac{dI}{dV}\delta V\cos(\omega t) + \frac{1}{2}\left(\frac{d^2I}{dV^2}\right)\delta V^2\cos^2(\omega t) + ... \qquad (10.6.4)$$

Here I_{DC} is the DC component of the current. Thus measuring the amplitude of the second term, which is at frequency ω, directly gives the differential conductance g. Measurement of the 2ω component gives the second derivative d^2I/dV^2.

A schematic of the electronic mixing circuit and the junction is shown in Figure 10.16 [34]. In general the basic idea can be realized in experiment in a number of ways. One realization can be that a function generator provides the DC bias in the form of a very low frequency triangular wave ramp ($f \sim 10$ mHz). The small amplitude *AC* signal $\delta V\cos(\omega t)$, obtained as the sine wave output of the lock-in amplifier (which is also the reference signal to the lock-in), is suitably attenuated so that $\delta V \approx 100 - 200\ \mu V$. An alternative is to have a dc and ac signal mixed by a synthesizer programme. Two op-amps maintain one end of the junction at a

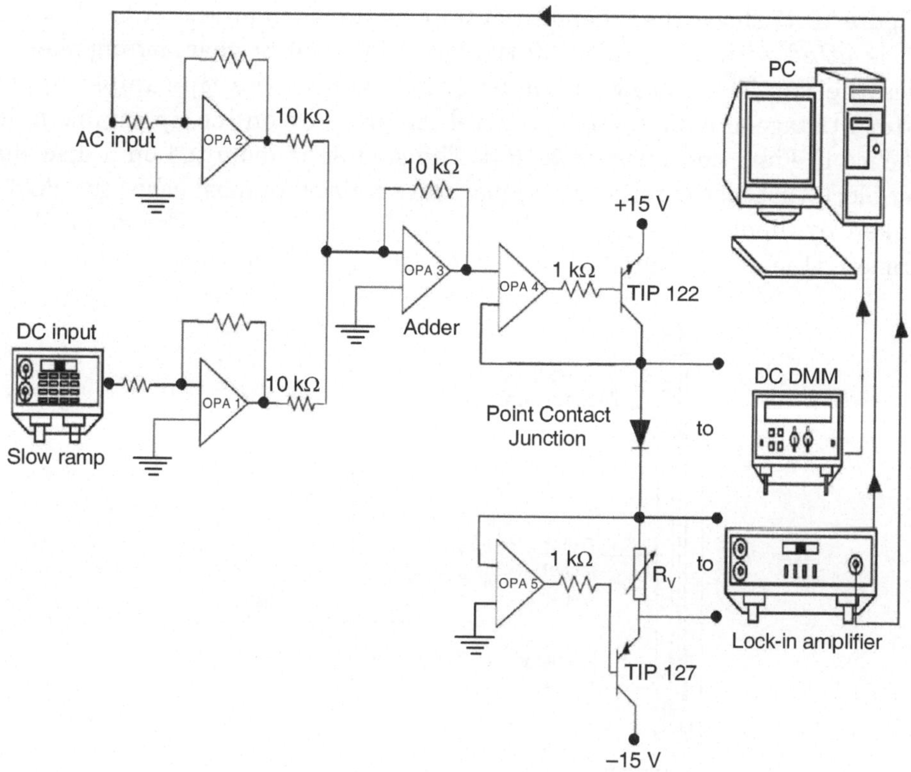

Figure 10.16: An electronic mixer circuit for PCS [2, 35]

potential $V_{DC} + \delta V\cos(\omega t)$ and the other end at a common ground potential. This is the most important feature of the circuit as the input modulated voltage should be kept constant across the junction irrespective of any changes in the junction resistance. For a given DC bias, the oscillations in δI (due to $\delta V\cos(\omega t)$) is measured by measuring the voltage across the resistor R_v (kept in series with the junction) by a lock-in amplifier (A-B mode). The value of R_v is adjusted close to the junction resistance. The rms value of the signal detected by the lock-in amplifier at the reference frequency (327 Hz) gives a measure of the dynamic conductance (dI/dV) of the junction. Transistors at the output of the op amps serve as current boosters. (These are necessary for measuring the dI/dV signal of low resistance junctions over a large voltage range as the op-amps can supply current only upto 10 mA). The DC bias voltage is monitored by connecting a DC voltmeter DMM across the junction. As V_{DC} is varied across the junction, we record dI/dV as a function of bias. A higher δV enhances the signal to noise ratio but at the same time will round off the features in the dI/dV within the energy range $\sim\delta V$. However the energy resolution is set by the step size ΔV_{DC}. The optimal value of δV is typically $(1/3)\Delta V_{DC}$.

Figure 10.17 shows the schematic of the cryostat used for the PCS. The cryostat insert is fitted with a precision 60 tpi screw in a worm gear arrangement that carries the tip. This screw arrangement serves as the tip-sample approach adjustment stage and allows us to control the pressure on the tip-sample junction and hence the junction resistance (R_J). The sample is mounted on a disc shaped copper block that also carries the temperature sensor. In most cases the PCS data are taken by dipping the sample directly in liquid He bath. For doing the PCS experiments the electrochemically etched Au tips were then plunged onto the sample

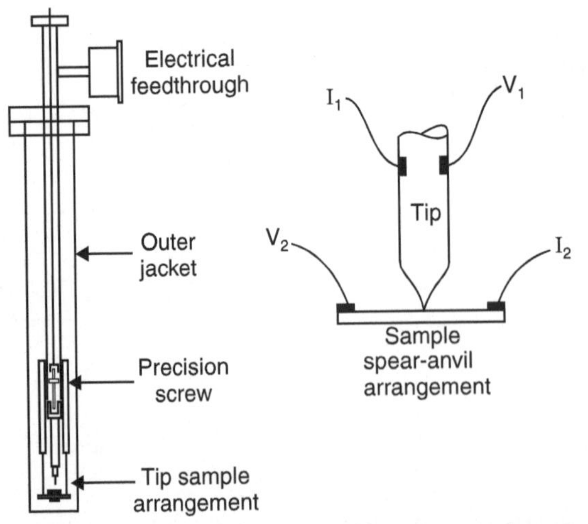

Figure 10.17: A cryostat for point-contact spectroscopy [35]

surface in the spear-anvil geometry. The tip was then retracted to vary the junction resistance. This procedure serves a dual purpose. Firstly Au tip under pressure starts 'flowing' which on retraction forms a sharper tip. Also the initial piercing of the sample surface leads to the exposure of a fresh sample surface that is highly desirable. Further the sample surfaces were freshly scraped and polished before loading them into the cryostat.

Example of point contact spectroscopy in the context of a complex oxide is shown in Figure 10.18 where we show the PCS data obtained on a single crystal of $Na_{0.9}WO_3$. This is a metallic oxide with cubic perovskite structure. For comparison in Figure 10.15 we show the PCS data taken on Au. In this material there is not much of spectral weight left in the PCS data beyond 20 meV. In contrast in the perovskite oxide there is a large spectral weight even at 90 meV. This spectral weight at higher energy is from the optical phonons.

One can use this PCS data to calculate the electron-phonon contribution to resistivity of this material using the relation [38, 39]:

$$\rho_{el-ph} = \frac{16\pi^2}{\omega_p^2 k_B T} \int_0^{\omega_{max}} \frac{\hbar\omega\alpha_{tr}^2 F(\omega)}{(e^{\hbar\omega/k_B T} - 1)(1 - e^{-\hbar\omega/k_B T})} d\omega \qquad (10.6.5)$$

The results are shown along with the experimental data on resistivity in Figure 10.19. One can see the excellent fit. In the calculation of the resistivity there is one fit parameter namely the Plasma frequency ω_p, which also can be compared with the experimental value. The example given above shows how a simple experiment can give useful information about a rather fundamental experimental quantity.

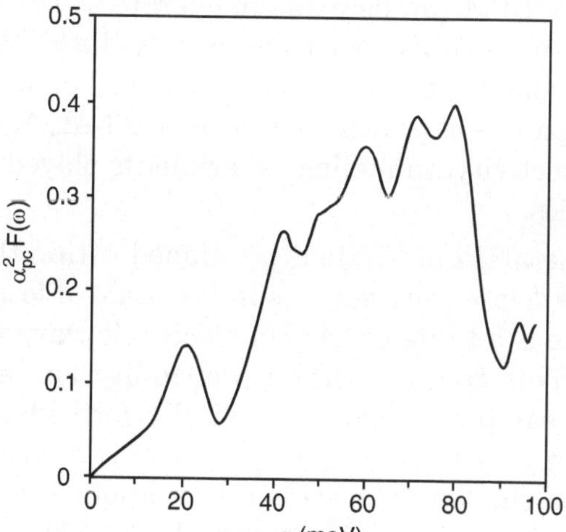

Figure 10.18: Point contact phonon spectrum of a metallic oxide Sodium Tungsten Bronze ($Na_{0.9}WO_3$) [37]

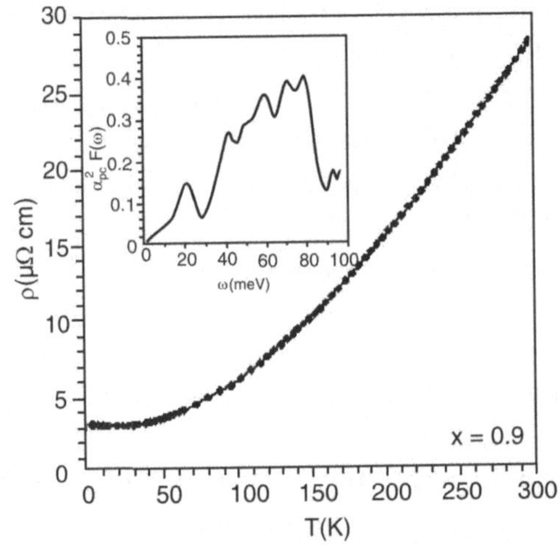

Figure 10.19: An example of the resistivity calculated from equation 10.6.5 based on the phonon spectrum (Eliashberg function) shown in the inset. [37]

7. SCANNING TUNNELING MICROSCOPY (STM) AND SCANNING TUNNELING SPECTROSCOPY (STS)

7.1. Basic Concept

In this section we discuss one of the most important developments of the last decade and a half, which added a very important tool in the field of solid state measurements. This is the field of STM and its different modifications [40]. The discovery of different modifications of STM and Atomic Force Microscope (AFM) is referred to as the field of Scanning Probe Microscopy (SPM) [41]. At present almost all the modifications of SPM can be made to operate at low temperatures. In this section we discuss the most widely used one-namely the STM and STS. It has to be noted that the phenomenon of tunneling has been used in low temperature physics for a long time through use of grown barrier as in Al-Al_2O_3-Al junction. In fact in the field of superconductivity, tunneling experiments played a very important role to elucidate the physics.

In STM a sharp metal (Pt or W) tip is positioned within the "close proximity" of a sample surface. The "close proximity" is in the scale of less than a nm (typically ~0.5 nm) so that there is a finite overlap of electronic wave functions of the sharp tip and the sample. This creates a finite probability for the electrons to tunnel between the tip and sample (which is classically forbidden). As a result of this finite overlap an electron can tunnel between the two electrodes. A measurable current flows between the two if there is a bias applied. Figure 10.20 shows the energy diagram of the tip and sample separated by a distance *s*. As *s* decreases to a few angstroms the tip and sample Fermi levels equalize resulting in equal and opposite current flow between them. On application of a positive bias, say to the sample, a net current starts flowing with the electrons from the occupied states of the tip tunneling to the empty states of the sample, as shown in Figure 10.20. The tunneling current I, at temperature T, is given by

$$I(s,V,W,T) = c\int_{-\infty}^{+\infty} N_S\left(E+\frac{eV}{2}\right)N_t\left(E-\frac{eV}{2}\right)\hat{t}(s,E,W)[f\left(E-\frac{eV}{2}\right)-f\left(E+\frac{eV}{2}\right)]dE \tag{10.7.1}$$

where c is a constant depending on the tip-sample effective function area, $N_t(E)$ and $N_s(E)$ are the tip and sample density of states (DOS), E is the energy of the tunneling electron, $f(E)$ is the Fermi function at temperature T, V is the applied bias between the tip and the sample. The expression above (eqn. 10.7.1) is symmetric in form. In our convention a positive bias implies that the sample is positive with respect to the tip. The transmission parameter t is the barrier penetration factor. s and W are the tip-sample distance and the average work function of the tip and sample surface respectively.

Usually the barrier is modeled as a one-dimensional trapezoidal barrier. In this case the barrier penetration factor (tunneling probability) t, can be calculated in the semi-classical WKB approximation. It is given as,

$$\hat{t}(s, E, W) \approx \exp(-2ks\sqrt{2(W-E)}) \tag{10.7.2}$$

where $k = m^{1/2}h$. In the above expression all the energies are measured with respect to E_F that is zero ofthe energy scale.

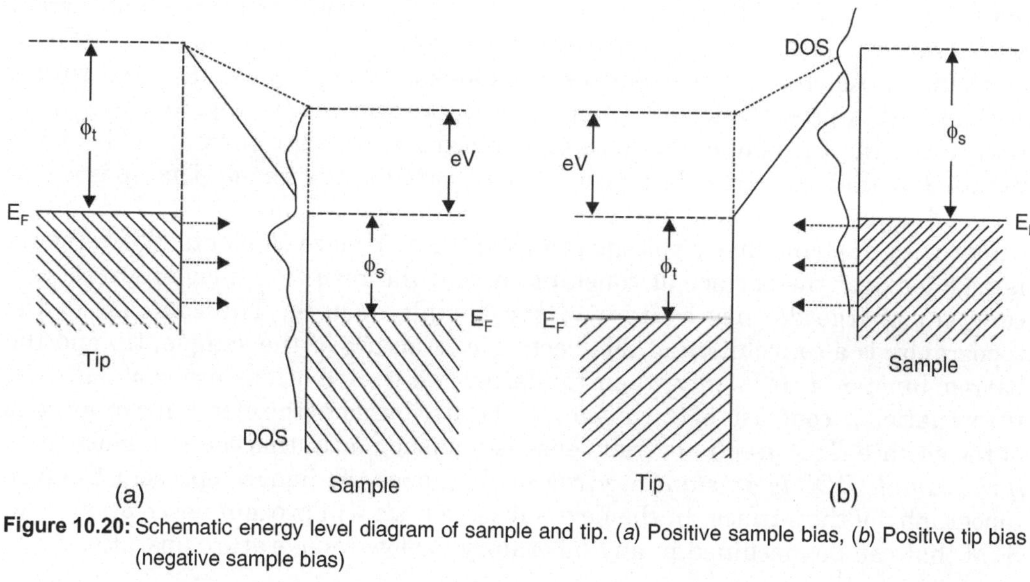

Figure 10.20: Schematic energy level diagram of sample and tip. (*a*) Positive sample bias, (*b*) Positive tip bias (negative sample bias)

Figure 10.21 shows the schematic of the STM. The operation of the STM depends on the crucial operation of placing the tip within a fraction of a nm over the sample and controlling this separation through a feed back loop. The other crucial operation is to obtain a scan that calls for sub-nm resolution. Both these have been made

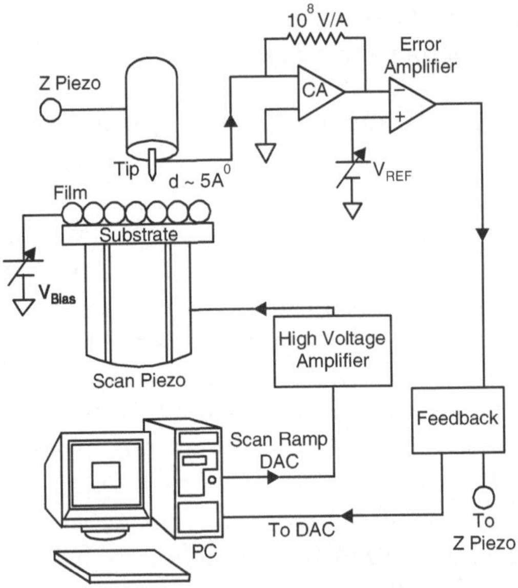

Figure 10.21: Schematic of the STM set-up [42]

possible by use of piezo-electric material like Pb-Zr-TiO_3 perovskite oxide. The tip is positioned over the sample surface by a piezo-electric tube (the Z piezo) while a constant bias is applied to the sample. The tip, maintained at a 'virtual' ground potential, acts as the electron source or sink depending on the sign of the bias applied to the sample. In a 'constant current' configuration, the tunneling current is kept constant as the tip is scanned over the sample. This is achieved by a feedback mechanism. As shown, the feedback mechanism compares the detected tunnel current with a preset value (V_{ref}) and gives a proportional voltage to the Z piezo that alters the tip-sample distance to maintain a constant current. This signal required to alter the vertical tip position is recorded as the image. The tip is raster scanned across the sample surface also by a piezo-positioner. This is generally achieved by applying linear voltage ramps to the X, Y piezo elements. When the tip is scanned over the surface at constant current the resulting image represents a constant charge density contour of the sample surface. The contrast in the topography is a convolution of the electronic structure of the sample, tip and the barrier function t. It is only when the latter two contributions are constant that the variation in contrast, of the image, can be attributed to the electronic properties of the surface. Topography in that sense is a misnomer in this context, since, *only if the sample DOS is constant over the surface*, does the image represent the true topography of the surface. In the next subsection we will present description of an STM that can be machined in any laboratory for low temperature operation.

7.2. Scanning Tunneling Spectroscopy (STS)

One of the major applications of STM at low temperatures is to do electron spectroscopy by the method of Scanning Tunneling Spectroscopy (STS). In STS, the STM based tunneling junction is used to probe the sample DOS as a function of the electron energy, much like the traditional tunneling spectroscopy. In STS the added advantage is to scan the surface and thus obtain spatially resolved information of the electronic properties. The technique involves positioning the tip over a point on the surface and establishing a constant tunnel current for a given DC bias and feedback set point. Now keeping the feedback voltage constant, so that the tip-sample distance remains unchanged, one 'sweeps' the sample bias to record the variation of I as a function of bias. Ideally a high sweep rate is desirable to avoid the effects of thermal drifts, inherent to the system. In practice the upper bound of the sweep rate is decided by the digitization speed (sampling rate) of the analog-digital converter that is used to record the data. In most available 16 bit D/A cards the speed is 200 KHz. Nyquist theorem thus puts a limit of around 100 KHz per channel. As an example of TS we show the tunneling data taken on a high T_c superconductor.

A schematic of the basic process of the STS is shown in Figure 10.20. When a positive bias is applied to the sample the dominant contribution to the tunneling current is from the unoccupied state of the sample and for negative bias the dominant contribution is from the occupied state. This can be qualitatively argued as follows. When a positive bias is applied to the sample, the sample energy levels are rigidly shifted downwards with respect to that of the tip and the net I arises from electrons tunneling from the occupied states of the tip to the unoccupied states of the sample (Figure 10.20). On the contrary if a positive bias is applied to the tip (equivalent to a negative sample bias) the tip and sample Fermi levels are displaced in the opposite direction. Now electrons from the occupied states of the sample tunnel to the empty states of the tip.

The total tunneling current is a convolution of the tip DOS $N_t(E)$ with the sample DOS $N_s(E)$ along with the transmission parameter t. The main objective of STS is to extract this sample DOS. The trick lies in the method of deconvolution of equation (10.7.1) to extract $N_s(E)$. If $N_t(E)$ does not have appreciable energy dependence in the energy range of investigation which is typically $|E| < \pm 1$ eV, the variation in the tunneling conductance with bias arises due to that of t and $N_t(E)$.

The details of exact deconvolution process can be found elsewhere [43, 44] Approximately if one assumes $N_t(E)$ to be energy independent and $N_s(E)$ to have an energy dependence slower than that of t then the tunnel conductance or dI/dV can be written as;

$$\frac{dI}{dV} = \frac{c}{2} N_t N_s \left\{ \hat{t}(s, \frac{eV}{2}, W) + \hat{t}(s, -\frac{eV}{2}, W) \right\} \tag{10.7.3}$$

When the DOS is nearly energy-independent as in a tunnel junction formed by noble metals, the energy (bias) dependence of dI/dV comes *entirely* from the barrier parameter t. As an example of STS data, Figure 10.22 shows the dI/dV data obtained on a Pt sample with Pt tip at 77 K. In this case we have $N_t(E) = N_s(E) \approx$ constant. The solid line in Figure 10.22 is the fit to equation (10.7.3). It can be seen that in this case the complete energy dependence comes from the barrier parameter.

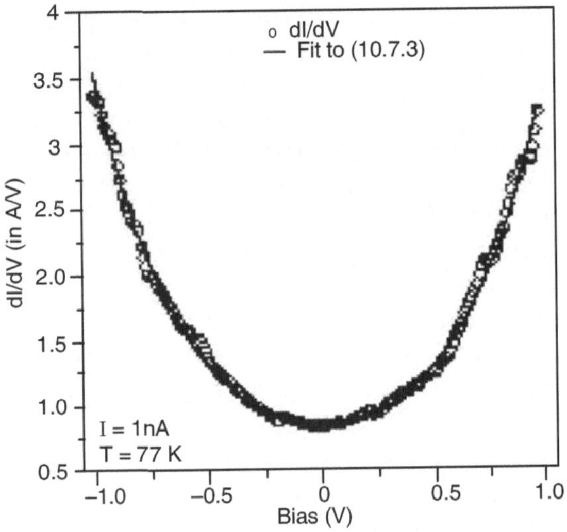

Figure 10.22: Tunneling Conductance as a function of bias for Pt-Pt vacuum tunnel junction by STS [45]

Normalizing the experimental data in general with the fit to the equation (10.7.3) leaves us with a quantity whose energy dependence reflects that of the DOS of tip and the sample and removes the barrier contribution. Using a Pt tip makes N_t(E) always energy independent. Thus such a normalization of *dI*/*dV* leaves us with the energy dependence of the sample DOS.

7.3. Map of Tunneling Conductance (C-MAP)

The STM also allows us to obtain the tunneling conductance *dI*/*dV* at a given bias *V* as a function of position. One can obtain the spectroscopic information simultaneously with the topographic image. In this case one can image tunneling conductance as a function of position. This is called "conductance map" (C-MAP). Conductance map is basically a 2*D* map of the tunnel conductance across the surface for a fixed bias and at a fixed average current. It is also possible, if there is a large memory, that one obtains the complete tunneling conductance (instead of at a given bias) over a range of bias at different positions of the substrate. One can get very useful information from such a map by making a Fourier transform. This has been applied to get information on quasi-particle scattering and Fermi surface in high T_c superconductors.

Since the tunneling conductance $dI/dV(V) \propto N_s(E = eV)$ the conductance map is a map of N_s. By a simple modification of the circuit one can have simultaneous recording of the topography and conductance maps. This is achieved by modulating the DC sample bias by a small AC signal ($V_{AC} << V_{DC}$) as has been done in the modulation method of PCS. The AC signal is at a frequency that is much larger than bandwidth of the STM electronics. Typically it is more than an order larger.

This makes the feed back circuit insensitive to the time varying ac signal. This allows us to detect the DC component of the tunneling current in a normal way by the current amplifier and is used to obtain the topography map in constant current mode. The AC component of I_t is measured by a lock-in amplifier with the in-phase component giving the $(dI/dV)_{Vdc}$. A very clear example of use of conductance map is in study of phase separation in rare-earth manganites showing CMR. These materials have an insulator to metal transition on cooling close to room temperature. This is also the ferromagnetic transition temperature of this material. Near the metal-insulator transition the material can undergo an electronic phase separation into an insulating phase and a conducting phase that grows on cooling. We show an example of this in Figure 10.23.

This is the C-MAP of an epitaxial film of $La_{0.67}Ca_{0.33}MnO_3$ (LCMO) grown on $SrTiO_3$ (STO). The film as shown in the Figure 10.23 has an insulator metal transition at T_p= 220 K. The film is strained uniformly and has a terrace and step growth with step height in the region of 0.4 nm. This can be seen in the topography image. The C-MAP taken at a temperature of 300 K does not have much contrast expect at the steps. But the C-MAP taken at 175 K (below T_p) shows contrast due to large variation of tunneling conductance at different locations of the film due to electronic phase separation.

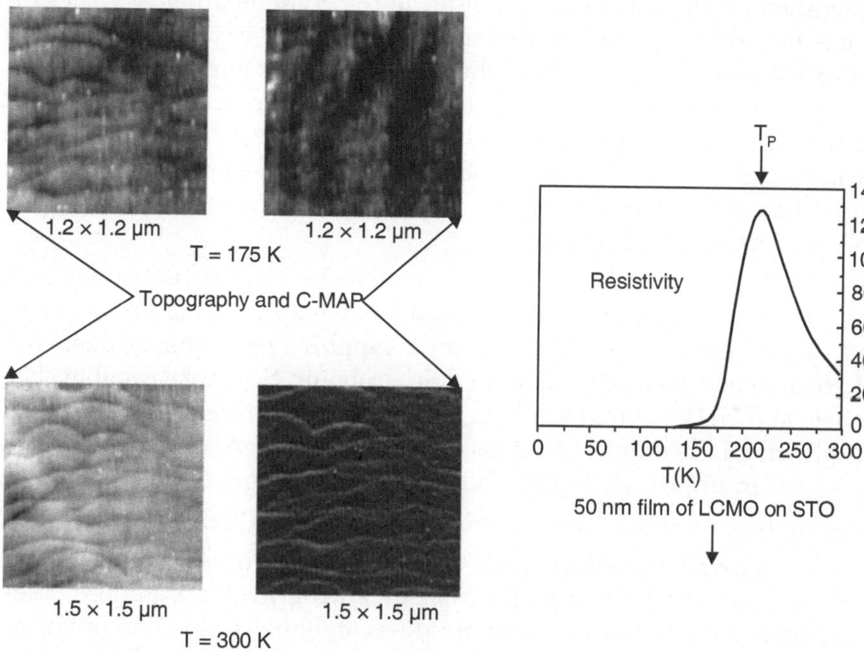

Figure 10.23: Simultaneous topography and C-MAP images of a film of LCMO grown on STO. The images are taken in scan range of 1.2-1.5 μm × 1.2-1.5 μm. The image at low temperature (T = 175 K) shows electronic phase separation. The right frame shows the resistivity of the film which has a T_p = 220 K [34]

7.4. Variable Temperature Scanning Tunneling Microscope

The acquisition of temperature dependent STM, STS as well as the C-MAP data can be done using a Variable temperature STM (VTSTM). A VTSTM which works in the range 4.2 K to 350 K can be constructed in the laboratory. It can also be made sufficiently small in size so that it can be placed in the space available inside the bore of a superconducting solenoid magnet. The main design considerations for the fabrication of the STM include an effective vibration isolation system, coarse positioner, scan unit, electronics and software for data acquisition and display of the scanned image. With little modifications, mainly in the electronics and software, one can use the STM for Scanning Tunneling Spectroscopy (STS), Scanning Tunneling Potentiometry (STP) and for generating conductance map (C-MAP) of the surface under study. The images shown in Figure 10.23 were taken by the VTSTM described below.

For vibration isolation the basic rule is to have the resonance frequency of the vibration isolation system far removed from the low frequency vibrations of the building and associated structural components. The natural frequency of the STM structure is typically much above a few KHz. If the STM is small this can be easily achieved. The building vibration is in the range of a few Hz to few tens of Hz. Thus the two mechanical systems can be made with widely different natural frequencies. Further isolation can be achieved by mounting the STM on a heavy structure that has very low natural frequency. This can be achieved by resting the cryostat that houses the STM on a heavy cast iron sheet supported by four air filled tubes. The air pressure in the tube is so adjusted that the natural frequency of the system is much below a Hz. This arrangement can be kept on a concrete reinforced foundation isolated from the rest of the flooring. This arrangement of a low resonant frequency of the support structure thus cuts down building vibration sufficiently to enable a reasonable resolution to be obtained.

A variety of electronic coarse positioners like Inchworm motor, piezoelectric louse, the beetle, inertial drive mechanism with magnetic clamping have been developed. A piezoelectric motor, based on a sapphire prism being pushed by six stacks of shear mode piezoelectric plates, is probably the most reliable. It is low temperature- and UHV-compatible. In a typical inertial drive (I-drive) mechanism, a slider is placed on the support that is connected to a piezo-electric tube. A simple design is shown in Figure 10.24 [46]. It consists of a 20 mm quartz tube (12.5 mm φ) attached to the end of the piezoelectric tube (length = 1″, diameter = 0.5″) using Stycast. The epoxy walker (slider), made of two halves of a brass cylinder, is inserted into the glass tube and is held in position by spring loaded stainless steel pins (located between the two halves). Two stainless steel balls glued on outer sides of the brass halves decrease the contact area between the tube and walker and prevent rolling motion. The correct spring tension is most critical since it directly controls the walker motion and its speed. A few trials and errors led to the successful adjustment of a spring. The *Z* piezoelectric tube carrying the tip is directly mounted on the walker. When a saw tooth wave is applied to the electrodes of the piezoelectric tube, the slow rise of the ramp elongates the piezoelectric tube carrying the quartz tube and walker along with it. In the sharp fall of the ramp the piezoelectric tube contracts leaving the walker in its position due to inertia. Hence there is a net motion of the walker with respect to the piezoelectric tube. While step size of the walker depends on the amplitude of the saw-tooth the net speed is dependent on the amplitude as well as the frequency. There are commercial sources from which such drives are available [47].

Figure 10.24 shows the schematic of the VTSTM. The head consists of three concentric Piezo-electric tubes (PET). The concentric design is found to minimize thermal drift and makes the STM head really compact. The outermost PET (called Scan piezo, Length = 2″, dia = 0.75″) is sectored into four symmetric quadrants along its length and used as the *XY* scan piezoelectric tube. Two diametrically

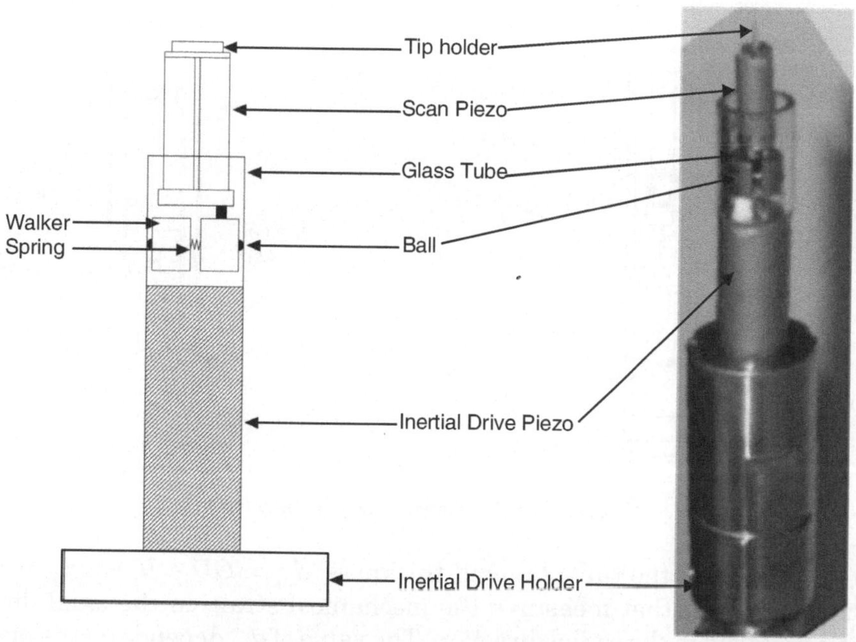

Figure 10.24: A simple piezo-electric walker [34]

opposite quadrants serve as *X* scan quadrants and the other two as the *Y* scan quadrants. The intermediate PET (Length = 1″, dia = 0.5″) is used for inertial drive as described before. The innermost PET (called *Z*-piezo, Length = 0.5″, dia = 0.25″) is used for fine approach and to control the tip-substrate separation. The inner electrodes of all the piezo tubes are held at ground potential. Electrical connections to the piezo electrodes were done using low melting point solders. Movement of the Piezoelectric tubes for the *X* and *Y* motion (which are sectored to achieve this motion) can be calculated from the formula:

$$\Delta X, \Delta Y = \frac{0.9 d_{31} L^3}{d_m t} V \tag{10.7.4}$$

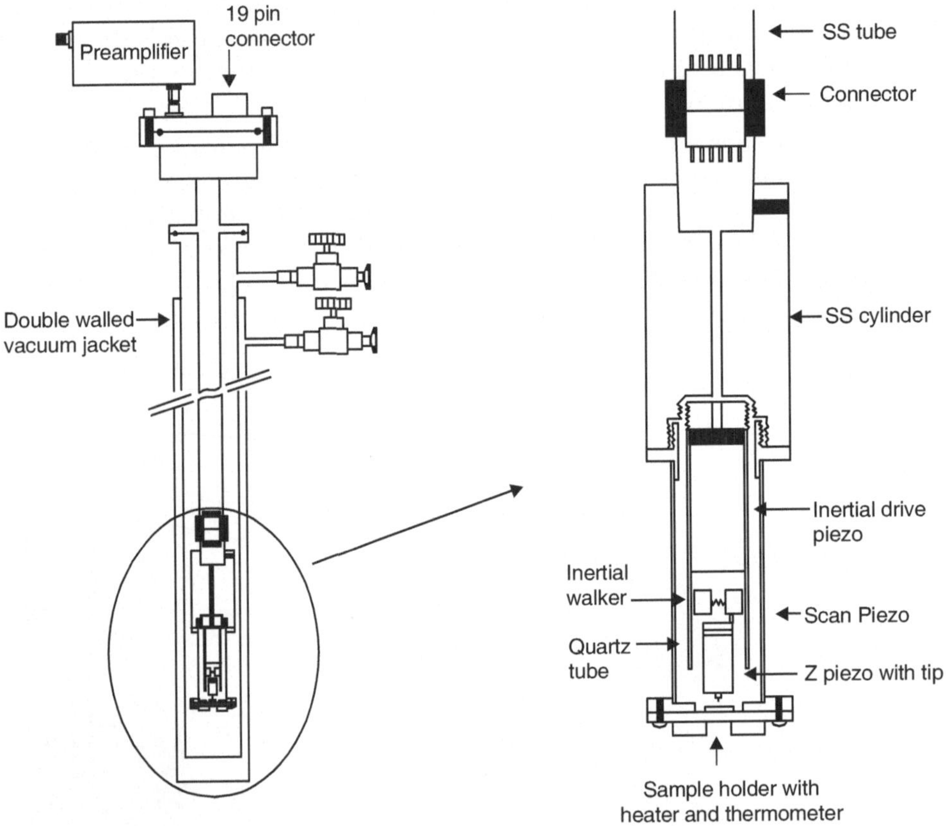

Figure 10.25: Schematic design of the VTSTM [42]

Here L = length of the tube, t = wall-thickness, $d_m = (OD + ID)/2$. d_{31} is the piezoelectric coefficient that measures the mechanical strain in the axial direction for unit electric field in the radial direction. The value of d_{31} depends on the piezoelectric ceramic material. Most of these materials are Pb-Zr-TiO_3 based ceramics. One of the problems in the low temperature use of these ceramics is that they lose sensitivity by nearly 25% down to 77 K and by 50% at He temperature. A proper calibration is often needed. For this HOPG graphite is often used as standard. Figure 10.26 shows the atomically resolved HOPG graphite image taken by the VTSTM.

The sample holder is kept at the head of the scan piezo. It should be made small; else the scan piezo will not be able to move it for scanning. A small heater (~25 W) and a calibrated thermometer can be mounted at the back of the sample holder for temperature control and measurements. The Z-piezo, mounted on the

I-drive walker carries the tip holder (a piece of hypodermic syringe needle) glued to the Z-piezo. One can use Pt/Rh(13%) wire (diameter of 0.1–0.2 mm) as a tip for the STM. It is advisable to use miniature coaxial cables for the tunnel current, bias and the feedback signals that need particularly good shielding to minimize noise. The coaxial cable carrying the tunnel current terminates in the current amplifier mounted on top of the cryostat. This is needed for the low noise; else long signal leads can pick up substantial noise. The described STM insert can be inserted into a double walled vacuum jacket for vacuum.

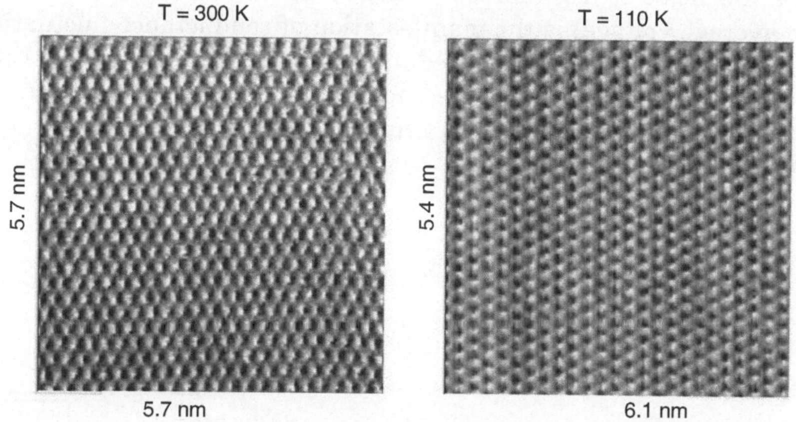

Figure 10.26: Images of HOPG graphite taken by the VTSTM [42]

The electronics is an extremely important factor which decides performance of the STM. While operating STM in the constant current mode, a variety of electronic circuits are required to amplify very small tunnel currents, to bias the sample, to control the tip sample distance, in order to maintain a preset tunnel current and to raster scan the tip across the sample. The main electronic blocks in a STM are the following: (*i*) Low noise preamplifier with a gain of 10^8 V/A, (*ii*) Bias circuit, (*iii*) Feedback circuit, (*iv*) Scan circuit and (*v*) High Voltage amplifiers to drive PET's.

For data acquisition and real time image display one can use a Pentium PC. Data acquisition is done by a 16-bit analog-digital (AD) conversion card. The card has piggy back digital-analog (DA) conversion module which generates the linear ramp voltages for the scan module. The image data acquired as an ASCII file can be processed using commercial softwares like SPIP^{TM} and softwares written in MATLAB^{TM}.

8. MEASUREMENT OF NOISE AND FLUCTUATIONS

Study of conductance fluctuation is a valuable tool to study low frequency dynamics in condensed matter systems. The conductance fluctuations in a current carrying conductor often show a power spectrum $S_V(f) = 1/f^{\alpha}$, $\alpha \approx 0.8–1.2$ and this is referred to as "1/*f* noise". A proper application of 1/*f* noise spectroscopy can be useful in different areas of materials science and condensed matter physics. In this chapter we would like to describe the experimental aspects for measurement of conductance or resistance fluctuations that often involve detection of voltage changes of the order of 0.5 nV–5 nV across the sample. Another purpose of describing these techniques is to have an example of digital signal processing which is often used in modern physical measurements. The sample (the resistor) is usually current biased by a low noise source and the voltage fluctuation ($\delta v(t) = v(t) - \langle v \rangle$, where $\langle v \rangle$ is the average voltage) is the manifestation of conductance fluctuation. $v(t)$ is the physical quantity that is measured and the measurement over a time period makes a time series. This is depicted schematically in Figure 10.27. The most common though not the only way to represent the noise is through its *spectral weight* $S_v(f)$ defined as:

$$S_v(f) = \underset{T\to\infty}{Lt}\left(\frac{1}{2T}\right)\left[\int_{-T}^{+T} dt\delta v(t)e^{-2ift}\right]^2 \tag{10.8.1}$$

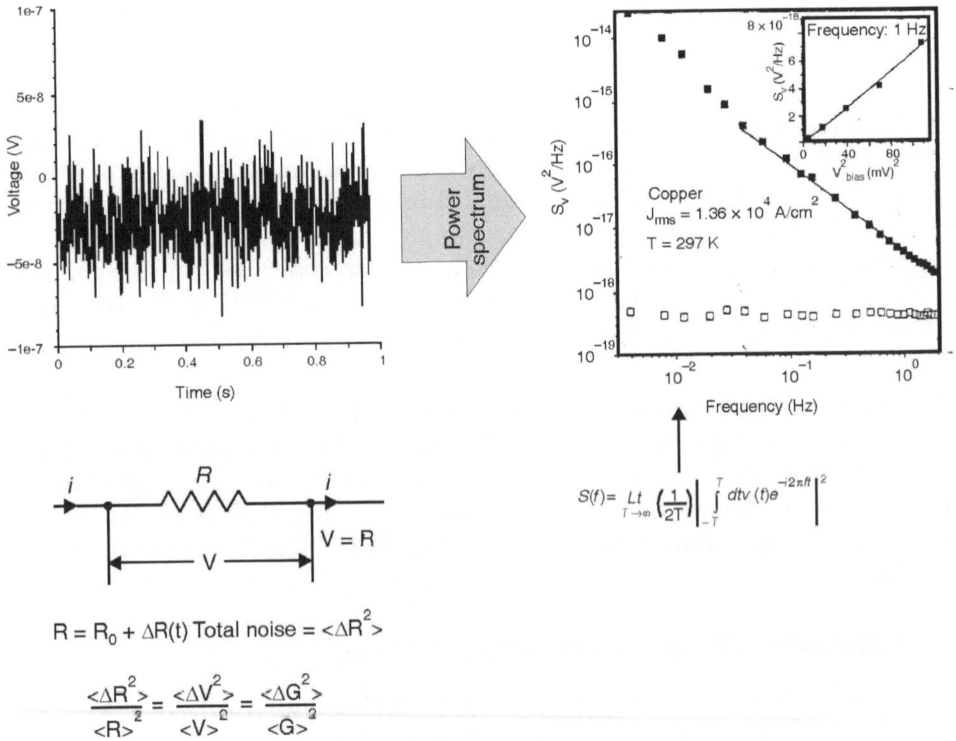

Figure 10.27: Basic schematic of noise measurement

Power spectrum is also the Fourier transform of the *auto-correlation function* $C(\tau)$ defined as:

$$C(\tau) \equiv \underset{T\to\infty}{Lt}\left(\frac{1}{2T}\right)\int_{-T}^{+T} dt\delta v(t+\tau)\delta v(t) \tag{10.8.2}$$

Often the noise is expressed as the rms fluctuation $<(\delta v^2)>^{1/2}$. This has to be defined in a band width of measurement between f_{max} and f_{min} so that

$$<(\delta v)^2> = \int_{f_{min}}^{f_{max}} df S_v(f) \tag{10.8.3}$$

The type of electrical noise that one sees in a solid are generally of three distinct types: (*a*) *Nyquist noise,* (*b*) *Shot noise and* (*c*) *Excess noise.*

Nyquist noise (often known as *Johnson noise* or Thermal noise) arises from the Nyquist theorem (discussed in chapter 8, see reference 4 of chapter 8). In a resistor of value R kept at a temperature T, it shows up as white noise ($S(f)$ independent of f) of spectral power:

$$S_v(f) = 4k_B TR \tag{10.8.4}$$

The Nyquist noise directly follows from the Fluctuation Dissipation Theorem (FDT) and $S_v(f)$ is independent of the current through the resistor and is not dependent on the materials property of the resistor. Consequently, it is not a useful tool of materials research. However it can be a measure of the effective electron temperature. In fact the Nyquist noise gives a way to measure the absolute temperature T.

The shot noise arises from discrete carrier motion in devices. This is also white in nature. However, it is proportional to the value of charge carrier (q) and the current (I) through the device. It has a spectral power density given as:

$$S_I(f) = 2qI \tag{10.8.5}$$

Shot noise is a good indicator of the magnitude of the charge carrier. In general, the shot noise is visible at low temperatures when the thermal noise is low enough.

The noise, which is of particular interest to most of the materials science applications, is the excess noise or the conductivity noise. This noise exists in a solid in addition to the thermal noise and shot noise. It is measurable when a given conductor is carrying a current. The spectral power of the voltage noise ($S_v(f)$) depends quadratically on the applied bias ($S_v(f) \propto v^2$). Due to the quadratic dependence of the spectral power on the bias, it increases and rises above the thermal noise only at a finite bias. This noise has a power spectrum $S_v(f) \propto f^{-\alpha}$. When $\alpha \approx 1$, this is referred to as the "1/*f* noise". Often the spectral power contains discrete or narrow band noise adding to the broad 1/*f* feature. Since the spectral power contains information on the dynamics of defect motion or relaxation, it is a very useful tool in material characterization [48, 49].

When one does noise spectroscopy it is important to realize that the noise from the sample is not the only source from which the noise arises. In fact the important task that one needs to do is to make sure that the extra noises are eliminated either by hardware or by software methods. Technique described here, though computational intensive, can be carried out in a modern Pentium PC. The technique described here can in general be used down to 1 mHz. The same technique with some modifications can be used over a wider frequency range [50].

The quantity observed experimentally is the fluctuating physical quantity (say $\delta v(t) \equiv v(t) - \langle v \rangle$, a voltage $v(t)$ as a function of time fluctuating around a mean $\langle v \rangle$). The main information is contained in this time series $\delta v(t)$. Often the information can be obtained from the distribution of voltage jumps $P(\delta v(t))$ or the rms fluctuation within a given bandwidth, $\langle(\delta v(t))^2\rangle^{1/2}$. However, the most popular way to represent the noise is through the power spectrum $S(f)$. The frequency range for which the spectral power is to be determined depends on the time scale of the fluctuations being investigated. In general, if the dynamics are occurring on a time scale τ, it will affect the power spectrum around the frequency range $f \sim (1/2\pi\tau)$. It is also important to ensure that the noise originates from the sample since many of the associated measuring electronics also have $1/f$ noise.

Noise voltage in a conductor can be measured by a simple dc method where a capacitor bank forming a high pass filter blocks off the dc voltage across the sample. The fluctuating voltage across the material to be studied is then amplified by a low-noise preamplifier (for samples with relatively higher resistance) or a low noise transformer for more metallic samples. The power spectrum can be obtained using a commercial spectrum analyzer or the time series, $\delta v(t)$, can be digitized and the power spectrum be obtained from the FFT of the correlation function of the time series. The ease of use has made the spectrum analyzer the most common method. However, in this method one loses all the information on the time series like the $P(\delta v(t))$. Since the $1/f$ noise is generally measured well below 10 KHz, digitizing the time series and using a separate FFT programme is a better alternative. This is because, the digitized time series $\delta v(t)$ is stored in the computer and one can carry out all the algebra with it leading to a more complete analysis of the data. With availability of relatively high speed (>200 KHz) A/D cards (more than 16-bits) and the decreasing cost of computers, this is definitely a preferred option. In order to separate out the $S(f)$ due to background the data are taken with zero bias current and then $S(f)$ is measured with a finite dc bias. The difference gives the noise due to the system under measurement.

The 4-probe dc method suffers from two very important drawbacks: (*a*) the $1/f$ noise of the preamplifier and the associated electronics adds to the background and raises the noise floor of the measurements and (*b*) the background estimation can be very faulty because it can shift during the measurement itself since long data collection times are necessary at lower frequency. If one uses an ac technique, both the drawbacks can be removed.

In the ac technique, the sample is biased by an ac and the signal is demodulated by a lock-in amplifier. This shifts the detection frequency to a region where the preamplifier has very low $1/f$ component. The signal from the lock-in is then digitized where the bandwidth of the low-pass filter of the lock-in amplifier will dictate the maximum frequency of the spectrum. In this case the carrier frequency $f >>$ bandwidth of measurement. For measuring spectra with $f < 10^2$ Hz this is a convenient technique. In addition, if the signal is recorded at two phases, it can be shown that the in-phase part will give a power spectrum that is a sum of both background and "signal" while the out-of-phase component gives the background only. The power spectrum of the noise from the sample ("signal") can be obtained by subtracting the background, which is measured simultaneously. This simultaneous measurement of the background is the greatest advantage of the ac technique. A much-improved version of the ac technique is the *5-probe technique* where the middle electrode is grounded. The sample forms two arms of a bridge, which can be balanced by low noise wire-wound resistors in the other two arms. Balancing the bridge can raise the sensitivity many fold and the contributions to noise by temperature drift etc. are largely cancelled out. Thus the 5-probe ac technique gives higher sensitivity [48, 50].

The methods to be used for noise measurements depend on the extent of noise which one would like to measure. When the spectral power of the sample noise is $>10^{-15}$ V^2/Hz, the background is often not a problem and the simple dc method will suffice. However, noise measurements in relatively clean systems and metallic systems can involve spectral power as low as 10^{-20} V^2/Hz or even lower. In that case the measurement requires special care and the background noise subtraction is a very important issue. The 5-probe ac method (using DSP techniques described in chapter 8) is the best solution for low noise measurements. This allows a spectral noise floor of lower than 10^{-20} V^2/Hz and allows measurements down to 1 mHz. The computational speed and power available in modern PC's can be effectively utilized for this purpose.

Noise measurements often suffer from the problem of irreproducibility, particularly in the absolute magnitude. There are two principal causes for this: (*a*) improper background subtraction and (*b*) contact noise. If the background is uncertain (as it often happens in a 4-probe dc method), then its subtraction may not be accurate enough, leading to irreproducibility of results. In an ideal case the background should be close to the thermal background of 4 k_BTR if the measurements are carried out in a shielded enclosure. In order to ensure that the background is reproducible it is good practice to check that the measured background noise is close to the expected thermal noise. The contact noise is another source of irreproducibility of noise measurements. Even in a 4-probe or 5-probe geometry one can get contributions from the contact noise which can itself have $1/f$ type spectral power. An ohmic contact or an observed quadratic dependence of $S(f)$ on the bias is no guarantee that the contact noise is negligible. The absence of contact noise contributions can be tested by making measurements with contact areas of different sizes and also by making measurements with different sample volumes (Ω). The measured $S(f)$ should follow the relation $S(f) \propto 1/\Omega$. An inverse volume dependence of the observed noise is the best indication that the measured noise is originating from the bulk of the solid.

By simultaneous measurement of background and the total power, (using a lock-in based technique) 5-probe a.c. measurement eliminates a good part of the problem. These techniques can also be used in magnetic field and variable temperature environment. In this technique the sample with 5 probes makes a bridge arrangement. The noise is detected as sideband of a carrier frequency and the lock-in is used as a demodulator. The band width over which the noise is detected is given by $f = \pm 1/\tau$, where τ is the outfilter time constant. The digitized signal is decimated, passed through digital anti-aliasing and low pass filter and the power spectrum is determined. The in-phase part gives the total power while the out of phase part gives the background like thermal noise that is bias independent. This is depicted in Figure 10.28, where we show a simple schematic.

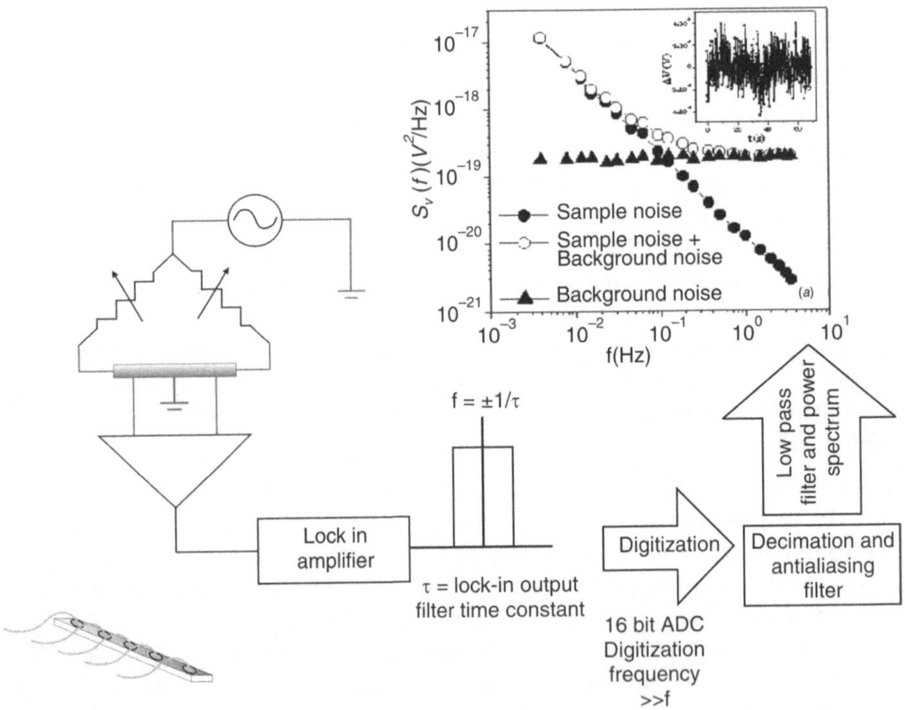

Figure 10.28: Schematic of 5 probe ac technique for noise determination [51]

9. CONCLUDING REMARKS

In this chapter we have described a number of techniques used in solid state physics measurements that can be used at low temperatures. We note that it is important to learn the basics involved in these measurements else the measurements will remain a black box.

REFERENCES

1. Rajeev K.P., Ph.D. Thesis Indian Institute of Science, Bangalore (1992),
Sangam Bannerjee Ph.D. Thesis Indian Institute of Science, Bangalore (1995).
2. ρ_0 H.H. Wieder, *"Laboratory notes on electrical and galvanomagnetic measurements",* (Elsevier) is a classical book covering a good description of methods. Some useful tips and modern realization can be found in Mandar A. Paranjape Ph.D. Thesis Indian Institute of Science, Bangalore (2004), Joy Mitra Ph.D. Thesis Indian Institute of Science, Bangalore (2004).
3. Ashcroft N.W. and Mermin N.D., "Solid State Physics"Holt-Saunders (1976);
MacDonald D.K.C., "Thermoelectricity an introduction to the principles", Wiley, New York, (1962).
4. Dugdale J., "Electrical Properties of Metals and alloys", Edward Arnold Ltd. London, (1977).
5. Eckland P.C. and Mabatah A.K., Rev. Sci. Instr., 48, 775, (1977);
Waclawek W., *et. al.,* J. Phys. E: Sci. Instrum., 14, 618, (1981);
Zhou Z. and Uher C., Rev. Sci. Instrum., 76, 023901, (2005).
6. Chen F., et. al., Rev. Sci. Instr., 72, 4201, (2001),
Howson M.A., Phys. Rev., B 47, 15324, (1993).
7. Rajeev K.P. and Vasanthacharya N.Y., Pramana, J. of Physics, 30, 29, (1988).
8. Mahendiran R., Tiwary S.K. and Raychaudhuri A.K., Solid state commn., 98, 701, (1996).
9. Banerjee S., Prins M.W.J., Rajeev K.P. and Raychaudhuri A.K., Pramana, J. Phys. 39, 391, (1992).
10. G. Vallebova, J. Thremal, Analysis, 16, 49, (1979); For a recent review on calorimetry see Kraftmakher Y., Phys. Rep. 356, 1 (2002) and "Comprehensive Handbook of Calorimetry and thermal analysis, "Edited by Japan Society of Calorimetry and thermal analysis, (English Edition, John Wiley).
11. Bachman D.R., et. al. Rev. Sci. Inst., 43, 205, (1972),
Griffing B.F. and Shivashankar S.A., Rev. Sci. Inst., 51, 1030, (1980),
Newsome R. and Andrei E.Y., Rev. Sci. Inst., 75, 104, (2004).
12. Carpentier L., et. al., J. Phys. D: Appl. Phys. 35, 402, (2000),
Kharkovbski H.I., Binek Ch., Kleenmann W., Appl. Phys. Letts., 77, 1, (2000),
Marcos J. et. al. Rev. Sci. Instr., 74, 4768, (2003),
Jin A.J. et. al., Rev. Sci. Instr. 69, 2439, (1998).
13. Rajeswari M., Ramasesha S.K. and Raychaudhuri A.K., J. Phys. E. Scientific Instr. 21, 1017 (1988).
14. Birge N.O. and Nagel S.R., Rev. Sci. Instr. 58, 1464, (1987).
15. Cahill D.J., Rev. Sci. Instr. 61, 802, (1990).
16. Sullivan P.F. and Siedel G., Phys. Rev. 173, 679, (1968).
17. Carslaw H.S. and Jaeger H.J., *"Heat Conduction in Solids",* (Oxford University Press) (1980).
18. Foner S., Rev. Sci. Instruments 27, 548, (1956) and ibid 30, 548, (1959).
19. Hoon S.R. and Willcock S.N.M., J. Phys E: Sci. Instr. 21, 772, (1988).
20. Krishnan R.V. and Bannerjee A., Rev. Sci. Instr. 70, 85, (1999).
21. Kingman R., Rowland S.C., and Popescu S., Am. J. Phys. 70, 595 (2002); see also the paper by Niazi A., Poddar P., and Rastogi A.K.,Current Science, 79, 99, (2000).
22. Rosenthal M.D. and Maxfield B.W., Rev. Sci. Instr. 46, 398, (1975),
Hockman A.J., Sena F.J. and Gentiloe P.S., Rev. Sci. Instr. 52, 224, (1981),
Deutz A.F., Hulstman R. and Kranenburg F.J., Rev. Sci. Instr. 60, 113, (1989),
Soumura T., Tsutsui K., Azuma T., Saito S., and Maeda T., Rev. Sci. Instr. 63, 4006, (1992).

23. A very complete description of the ac susceptibility measurements with examples can be found in the contribution by Sue Kilcoyne, Department of Physics and Astronomy, University of Leeds, Leeds LS2 9JT, available at http:conferences.iop.org/LAB/sue_ac_susc_final.ppt.
24. Hartshorn L., J. Sci. Instr. 2, 145, (1925).
25. Prasad M., Rao R.R. and Raychaudhuri A.K., J. Phys E: Sci. Instr. 19, 1013, (1986).
26. Chambers R.G. and Park J.G., Br. J. Appl. Phys. 12, 507, (1961).
27. Delaney J.A. and Pippard A.B., "Electrodeless methds for conductivity measurement in metals," Rep. Prog. Phys. 35, 677, (1972).
28. Kraftmakher Y., Meas. Sci. Technol. 2, 253, (1991), Bauhofer W., J. Phys. E: Sci. Instr. 10, 1212, (1977).
29. Edgar A. and Quilty J.W., Am. J. Phys. 61, 943, (1993).
30. Landau L.D., E.M. Lifshitz and L.P. Pitaevskii, "Electrodynamics of Continuous Media" (Pergamon, Oxford), (1984).
31. Yanson I.K., Kulik I.O. and Batrak A.G., Jl. Low Temp. Phys. 42, 527, (1981) and Yanson I.K. Soviet Phys. JETP 39, 506, (1974).
32. Diuf A.M., Janson A.G.M. and Wyder P., J. Phys.: Cond. Matter 1, 3157, (1989).
33. Andreev A.F., Sov. Phys., JETP 19, 1228, (1964) and Blonder G.E., Tinkham M. and Klapwijk T.M., Phys. Rev, B 25, 4515, (1980).
34. Srikanth H. and Raychaudhuri A.K., Phys. Rev. B 45, 383, (1991) and Phys. Rev., B 46, 1471, (1992).
35. Srikanth H. and Raychaudhuri A.K., Cryogenics, 31, 421, (1991).
36. Srikanth H. Ph.D. Thesis Indian Institute of Science, Bangalore, (1992).
37. Gayathri N. and Raychaudhuri A.K., J. Low Temperature Physics, 137, 471, (2004).
38. G. Grimvall, "The electron-phonon interaction in metals", North-Holland, (1984).
39. Allen P.B., Phys. Rev., B 34, 4331, (1986).
40. Binnig G., Rohrer H. and Weibel E., Phys. Rev. Lett. 49, 57, (1982).
41. J.A. Stroscio and W.J. Kaiser, "Scanning Tunneling Microscopy", Academic Press, San Diego, (1993).
42. A simple design of the STM set-up is given in ref. 2.
43. E.L. Wolf, "Principles of Electron Tunneling Spectroscopy", Oxford University Press, New York, (1985).
44. Vladimir A. Ukraintsev, Phys. Rev., B 53, 11 176, (1996).
45. Biswas A., Elizabeth S., Raychaudhuri A.K., and Bhat H.L., Phys. Rev. B 59, 5368, (1999).
46. Agrait, N., Rev. Sci. Instrum., 63, 263, (1992), Brar L.K., Mandar Pranjape, Ayan Guha and Raychaudhuri A.K., Current Science, 83, 1199, (2002).
47. There are low temperature inertial drives available from commercial sources. For example—www.attocube® source.
48. Raychaudhuri A.K., Current Opinion in Solid State & Materials Science 60, 67, (2002).
49. Dutta P. and Horn P.M., Rev. Mod. Phys. 53, 497, (1981),
Weissman M.B., Rev. Mod. Phys. 60, 537, (1988),
Jones B.K., Adv., in electronics and electron Physics, 87, 201, (1994).
50. References for measurements are given in reference 48. It contains also a discussion on different techniques used and their reference. For 5-probe ac measurement in particular see Scofield J.H., Rev. Sci. Instr. 58, 985, (1987).
51. Arindam Ghosh, Ph.D. Thesis, Indian Institute of Science, Bangalore (1999). A complete description of the measurement procedure can be found in Arindam Ghoch, Swastik Kar, Aveek Bid and Raychaudhuri A.K. arXiv:cond-mat/0402130 v. 1, 4, Feb. 2004.

APPENDIX

Elements of Vacuum Techniques

1. INTRODUCTION

In any low temperature experiment, the sample that is cooled is enclosed within a chamber. The walls of the chamber are at room temperature. In order to prevent transfer of heat from the room temperature walls to the cold sample through convection of heat, the chamber will have to be maintained at a pressure of 10^{-5} millibars or lower. One needs a basic understanding of vacuum techniques to carry out experiments successfully at low temperature. The range of pressure from 10^{-2} to 10^{-6} millibar is called high vacuum. The range of pressure below 10^{-6} millibar is called ultrahigh vacuum. Common experiments at low temperature are carried out in the high vacuum range. In this chapter some essential information on vacuum techniques in the high vacuum range will be given. For a deeper and a more extensive understanding of the subject one will have to refer to standard texts on vacuum techniques. Titles of a few such texts are given at the end of the chapter.

2. THROUGHPUT OF A GAS

A gas, especially at low pressure, obeys the Charles law

$$Pv = n\, k_B T \tag{A.2.1}$$

Here P is the pressure in Pascals[1], v is the volume in m^3, n is the number of molecules of the gas in the volume v, T is the temperature in degrees Kelvin and k_B is the Boltzmann constant, which has the value 1.381×10^{-23} Joule/K. At 300 K and 1 atmosphere (approximately 1000 mbar) pressure the number density, N, of molecules is 2.45×10^{25} per cubic metre. At a pressure of 10^{-6} mbar, the number density of molecules is $2.45 \times 10^{16}/m^3$.

Let us have a vessel of volume V litres. When the gas in the vessel is pumped out the pressure in the vessel decreases. Let the rate of decreases of pressure be δp millibar per second. Then the number of gas molecules removed per second at 300 K is

$$dn/dt = 2.45 \times 10^{19}\, V\, \delta p \tag{A.2.2}$$

The quantity $Q = V\,\delta p$ is called the throughput of the gas from the chamber. It is measured in standard millibar-litre per second.

---

[1] One Pascal is one Newton/m^2. Atmospheric pressure is 1.013×10^5 Pascals. Atmospheric pressure is referred to as one bar. A millibar is approximately 100 Pa. Pressure is measured in millibars in vacuum practice.

The word standard refers to the throughput when the vessel is at 273.16 K. Since the number density is inversely proportional to the absolute temperature, one standard millibar-litre/sec corresponds to the flow of 2.20×10^{19} molecules per second. We shall see later that this throughput determines the pressure drop along a line through which the gas is pumped.

3. MEAN FREE PATH OF A MOLECULE

There are a large number of molecules in a chamber and the molecules are moving in random directions with a distribution of velocity determined by the absolute temperature of the gas. The molecules are not point objects. They behave as objects which have a cross-sectional area σ. If such a molecule has a speed v, it sweeps out a volume $\sigma v dt$ in a time dt. It collides with all the molecules whose centres lie within this volume element. The number of molecules with which it collides in time dt is therefore $N\sigma v dt$. The time τ between two collisions is

$$\tau = 1/N\sigma v \tag{A.3.1}$$

The average distance the molecule travels between two successive collisions is called the mean free path l. The mean free path l is therefore given by

$$l = 1/N\sigma \tag{A.3.2}$$

Since N, the number density of molecules, varies inversely as the pressure, the mean free path l of a gas molecule varies inversely as the pressure. For a nitrogen molecule assume a collision cross-section of 6×10^{-21} m^2. At 300 K, at one bar pressure, the mean free path calculated from equation (A.3.2) is 6.8×10^{-6} m. This increases to 0.68 cm at a pressure of 1 millibar, 680 cm at 10^{-3} millibar, 6.8×10^{3} m at 10^{-6} m bar and so on.

For a given diameter of the channel there is a critical velocity at which the flow changes from non-turbulent to turbulent. The flow characteristics will be different for a turbulent and a non-turbulent flow. The *Reynolds number* for the flow is defined by

$$Re = dv\rho/\eta \tag{A.3.3}$$

Here d is an effective diameter for the channel, v is the velocity of flow, ρ is the density of the gas and η is its viscosity. For a tube of circular cross-section the actual diameter is the effective diameter. If Re is more than 2200 then the flow is turbulent. If it is less than 2200 the flow is non-turbulent.

Using the values for the density and viscosity of air at 20°C, a Reynolds number of 2200 corresponds to a velocity of about 70 cm/s. Usually in vacuum applications the flow velocity of air is less than this critical velocity and so we generally do not have to deal with turbulent flow in vacuum applications.

In non-turbulent flow we can distinguish between two extreme regimes of flow. In the first regime the mean free path of the gas molecule is much smaller than the dimensions of the channel. In this regime the flow is dominated by collisions between the molecules and so this regime is called the *viscous flow* regime. Such a regime is found at pressures in the range 1 bar to 10^{-2} mbar.

At the other extreme is the regime of flow in which the mean free path of the gas molecules is very much larger than the dimensions of the channel. In this regime intermolecular collisions are not important. The flow properties are governed by the collisions of the molecule with the walls of the channel. This regime is called *the molecular flow* regime.

For vessel diameters that are commonly used, such a regime is found at a pressure of 10^{-4} mbar or less. In the range of pressure from 10^{-2} to 10^{-4} mbar both intermolecular collisions and collisions of the molecules with the walls of the channel are important. This regime is called the *mixed flow* regime. For a given throughput the pressure drop across the channel will be governed by different relations in these regimes.

4. CONDUCTANCE

A pressure difference δp across a channel generates a flow of a gas through it. The flow is proportional to the pressure difference δp. We may define the *conductance* C of a channel by the relation

$$Q = C\,\delta p \tag{A.4.1}$$

The throughput is in std. millibar-litre per second, the pressure drop is in millibars and so the conductance C is measured in litres per sec. The conductance of a channel depends on the dimensions of the channel and the regime of flow through the channel. In Table A.1 are collected expressions for the conductance of an orifice and a circular pipe both in the viscous flow regime and in the molecular flow regime.

Table A.1: Expressions for the conductance for different types of channels in the viscous and molecular flow regimes. The dimensions of the channels are in cm.

Type of channel	*Conductance Expression for air at 20° C*		*Remarks*
	Viscous flow	*Molecular flow*	
Circular Orifice of radius r in a circular tube of radius R	$\dfrac{20\,r^2}{(1-P_1/P_0)}$ (*a*)	$\dfrac{1}{\alpha KA(1-r^2/R^2)}$ (*b*)	(*a*) P_0 is the pressure on the source side and $P_1/P_0 < 0.525$. (*b*) $K = 11.6$ *l*/s for air at 20°C. α varies from 1 to 1.33 as r increases from 0 to R.
Cylindrical pipe of Length l and diameter d	$\dfrac{2190\,p_{av}r^4}{l}$ when $p_{av}d > 0.72$ mbar.cm $\dfrac{8r^2(137\,dp_{av}+12.17)}{l}$ when $0.03 < p_{av} < 0.72$	$\dfrac{62\,F^2}{Ul}$ (*c*)	(*c*) U is the perimeter of the cross-section of the tube and F is the area of cross-section.

If N channels of conductance C_j ($j = 1 \ldots N$) are connected in series the effective conductance is given by

$$1/C_{eff} = \sum_{1}^{N}(1/C_j) \tag{A.4.2}$$

This is because the throughput through every channel is the same while the pressure drops across the channels add up. However this simple relation is not generally valid as the Maxwellian distribution of velocities is disturbed during pumping. In a long channel the distribution is distorted from the random direction of velocities at the chamber end, to a distribution in which there are more molecules in the forward direction at the pumping end.

Since the chamber to be evacuated will be connected to the pump through several components like bellows, tees, elbows and valves, there will be a considerable pressure drop between the outlet of the chamber and inlet of the pump, the pressure at the outlet of the chamber being higher. To achieve a vacuum in the chamber close to the ultimate vacuum achievable by the pump, the effective conductance of the pump line must be made large. This will also reduce the pump down time of the chamber from atmospheric pressure to the final low pressure.

5. PUMPS

To remove air in the chamber and reduce the pressure inside it one can use different types of pumps. A rotary vane pump will pump down from atmospheric pressure to about 10^{-3} mbar. The rotary vane pump is always used as the first stage in a pumping system. The rotary vane pump can be a single or double stage pump. A spring-loaded vane rotates at a speed of 200 to 500 rpm in a cylindrical casing (see Figure A.1).

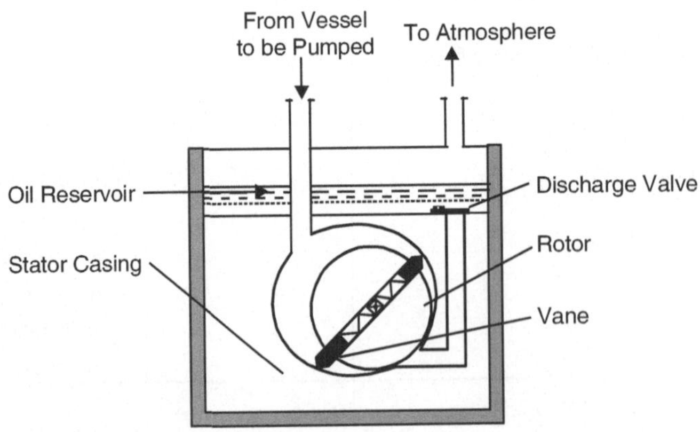

Figure A.1: A single stage rotary vane pump

The rotor is mounted slightly eccentrically with respect to the axis of the casing. The rotating vane removes the gas at low pressure from the chamber, compresses it to a pressure slightly above atmospheric pressure and discharges it to the atmosphere. In these pumps a gas ballast arrangement will be provided to remove condensable vapours. The oil in a rotary vane pump is used to lubricate the moving parts and to seal gas into compressible volumes. It also serves to transport heat away from wear surfaces and inhibits corrosion. Any vacuum-grade distilled hydrocarbon oil can be used. The only requirements are a low viscosity, and low vapour pressure. The hydrocarbon oils are made by distlling crude oil. These have a vapour pressure lower than 10^{-6} mbar. They are of low cost and low toxicity. More expensive synthetic oils such as silicones, poly-ethers and esters are used to obtain a better performance.

The speed of any pump is designated as *S* l/s. For rotary pumps the speed is designated in m^3/minute. Mechanical pumps are available in the speed range 0.05 m^3/minute to 3 m^3/minute. The rotary vane pump has been developed to a high degree of perfection to give noiseless vibration-free operation.

To get a pressure of 10^{-6} mbar that is sufficient for many low temperature applications and measurements, the rotary vane pump is followed by an oil diffusion pump or a turbo-molecular pump. In an oil diffusion pump (Figure A.2) a jet of hot oil vapour is generated which ejects and traps the molecules of air from the chamber to be evacuated.

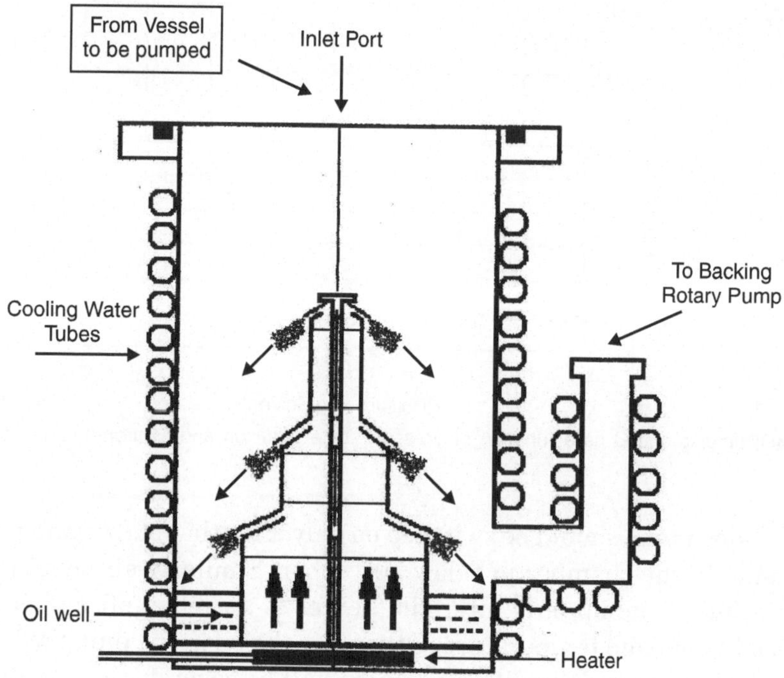

Figure A.2: Oil Diffusion Pump

[2]A pumping speed of *S* m^3 per minute corresponds to a pumping speed of 16.7 × *S* litres/s.

The oil is condensed at the pump wall, cooled by water or air, and flows back to the boiler. The ejected air is compressed, and it is sucked away by the rotary pump connected to the inlet port of the diffusion pump. The pumping speed of a diffusion pump is a function of the inlet pressure. It rises from a low value at an inlet pressure of 10^{-3} mbar, reaches its design steady value at a pressure below 10^{-7} mbar. As the pressure is lowered further the pumping speed drops. The saturated vapour pressure of the oil used sets the limiting value of the pressure ultimately reached. The saturated vapour pressure is usually in the range 10^{-9} mbar or lower.

The oil used in a diffusion pump should have a high molecular weight relative to air to increase the efficiency of ejection of air, should have a low viscosity and should be thermally stable against decomposition. Silicone oils which have a vapour pressure around 10^{-10} mbar are normally used. Diffusion pumps come with a range of pumping speeds from about 50 l/s to 100,000 l/s. The higher the pumping speed, the larger is the size of the pump as well as the size of the inlet port.

If we want to avoid back-streaming of the oil vapour from contaminating the chamber, we may include a baffle between the inlet port of the diffusion pump and the inlet of the chamber. The baffle can be cooled with water.

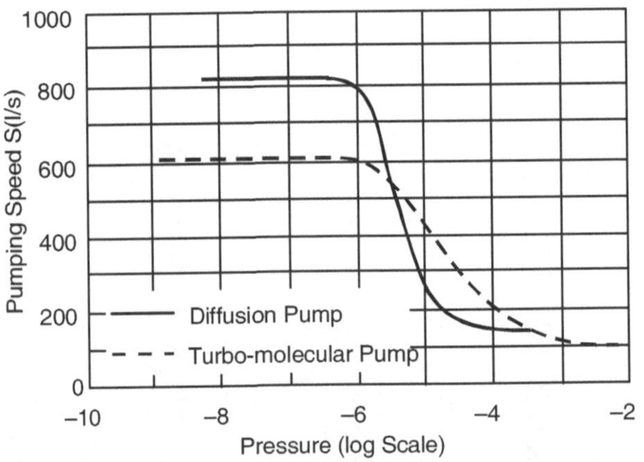

Figure A.3: Pumping speed as a function of pressure for a diffusion and a turbo-pump. The pressure is in mbars.

A diffusion pump should be switched on only after the rotary vane pump reduces the pressure in the chamber to below 10^{-2} mbar. Similarly air should be admitted into the diffusion pump only after the heater is switched off and the pump has cooled to a low enough temperature. Otherwise the oil in the pump will get oxidised and the performance of the diffusion pump will be severely degraded.

The turbo-molecular pump is a high tech alternative to the diffusion pump. If one wants a completely oil free high or ultrahigh vacuum, the turbo-molecular pump is to be preferred. For the same pumping speed the turbo-pump is much more expensive than the diffusion pump. Figure A.4 shows a schematic diagram of a turbo-molecular pump.

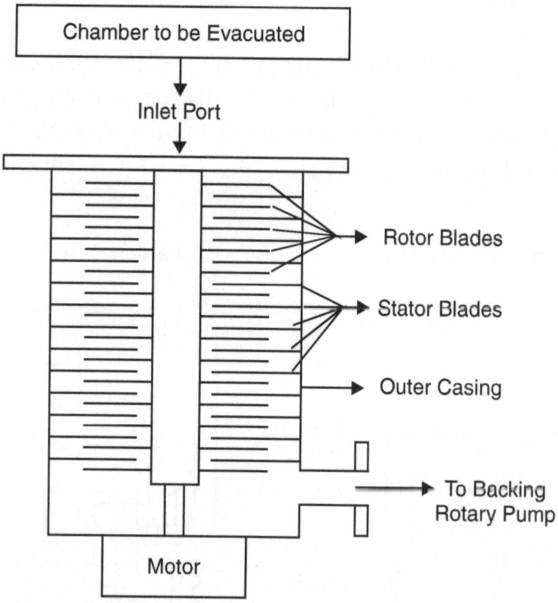

Figure A.4: Turbo-molecular Pump

In the turbo pump one has a set of rotor blades rotating at a high speed with respect to a set of stator blades. The gas molecules colliding with the rotating blades are pushed out. The speed of rotation is very high. One uses a dry lubricated ceramic bearing or a grease lubricated bearing for the turbine. After the rotary vane pump has reached a vacuum of 10^{-2} millibar or lower the turbo pump is switched on. It takes some time to reach its full speed of rotation. Then its pumping speed reaches a steady maximum value.

The pumping speed of a turbo-pump as a function of pressure is shown in Figure A.3. If the turbine is damaged it is now possible to get spare turbines which can be replaced in the grease lubricated bearing pumps by the user himself. If it is a dry lubricated pump the replacement of the bearing is a specialised job.

For certain special applications where one needs a very clean ultrahigh vacuum one uses an ion pump. In an ion pump the gas molecules are ionised by electrons accelerated to a few kilovolts. The positively charged ions are then collected on a negatively charged plate. The ion current is proportional to the number density of gas molecules and hence is an indication of the pressure. One can isolate the turbo-cum-rotary pump with a gate valve and switch them off, and use the ion pump alone to reach a vacuum of 10^{-8} millibar or lower.

6. GAUGES

We shall only mention two types of gauges which are commonly used in high vacuum practice. In the thermocouple gauge the temperature, or resistance, of a hot filament in the vacuum is measured. When the pressure in the system is 10^{-3} mbar or higher, the temperature of a hot filament will depend on the pressure of the gas since heat will be conducted away from the hot filament by the gas molecules colliding with it. The temperature of the hot filament or the resistance of the hot filament, which is a function of its temperature, can be measured for a constant current through the filament and calibrated against the pressure. The thermocouple gauge is shown in Figure A.5.

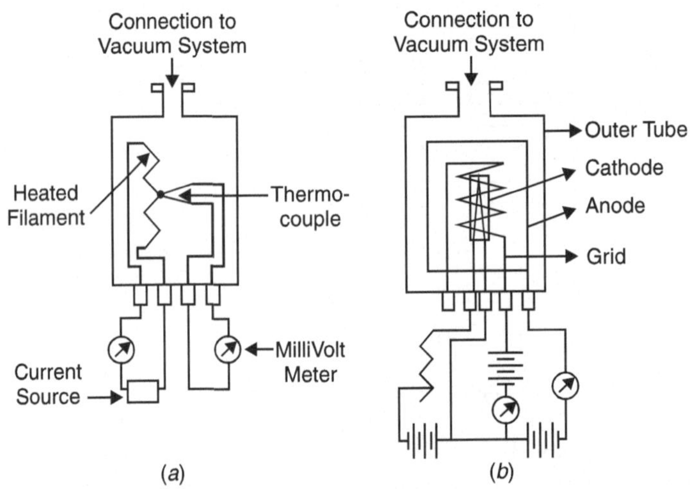

Figure A.5: (*a*) Thermocouple gauge; (*b*) Ionisation gauge

The thermocouple gauge is connected between the rotary pump and the diffusion or turbo-molecular pump. When the pressure on this gauge goes below 10^{-2} millibar the diffusion or turbo-molecular pump will be started. The thermocouple gauge cannot be used below 10^{-3} millibar.

For pressures below 10^{-4} millibar one uses an ionisation gauge. In this gauge electrons are accelerated between a hot filament and a grid that is kept at a positive potential with respect to the filament. The accelerated electrons ionise the gas molecules. The positive ions are collected in a cylindrical collector maintained at a slightly negative voltage with respect to the grid. The electron current is regulated at a steady value. The ion current is then directly proportional to the pressure.

However there could be photoelectron emission from the collector due to photons arising from the de-excitation of the ions or due to photons emitted when the electrons hit the grid. This photoelectric current from the collector plate sets a limit to the lowest pressure that can be measured by the ionisation gauge.

To measure pressures below 10^{-10} mbar one uses the *Bayard Alpert gauge*. Here the filament is on the outside and the collector is a thin wire along the axis of the grid. Since the area of surface of the collector is enormously reduced in this arrangement, the photoelectric current is reduced correspondingly.

The ion gauge heads can be purchased as nude gauge heads. In these gauge heads the electrode structure is directly mounted on a flange without any housing. The nude gauge head can be fixed to a mating flange on the vacuum system. The ion gauge should be switched on only when the vacuum has reached below 10^{-3} millibar on the thermocouple gauge.

7. VACUUM COUPLINGS

Several components make a vacuum system. We need a coupling at the location where two vacuum components are joined. The KF or NW vacuum coupling is a standard simple and flexible coupling down to 10^{-6} mbar. It consists of two flat flanges with a *centring ring*. The O-ring is on the centring ring. The centring ring is clamped between the two flanges using a clamp. Figure A.6 shows such a NW flange and Table A.2 gives the dimensions for various sizes in inches.

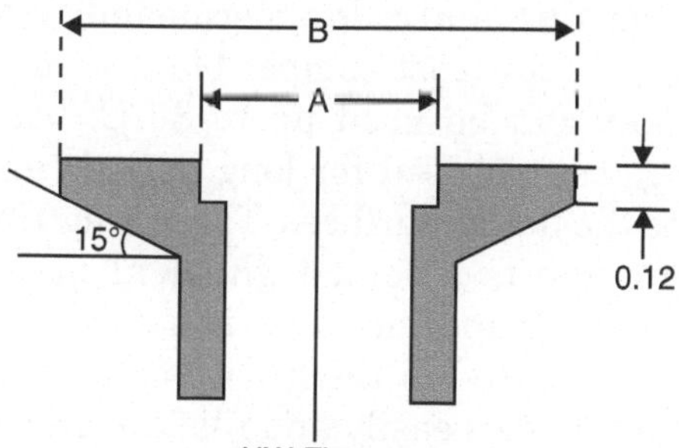

NW Flange
All dimensions in inches

Table A.2: All dimensions in inches

	NW10	*NW16*	*NW25*	*NW40*	*NW50*
A	0.48	0.68	1.03	1.62	2.06
B	1.18	1.18	1.57	2.16	2.95
Tube Size	0.50	0.75	1.0	1.5	2.0

Figure A.6 and Table A.2: NW type flanges along with dimensions

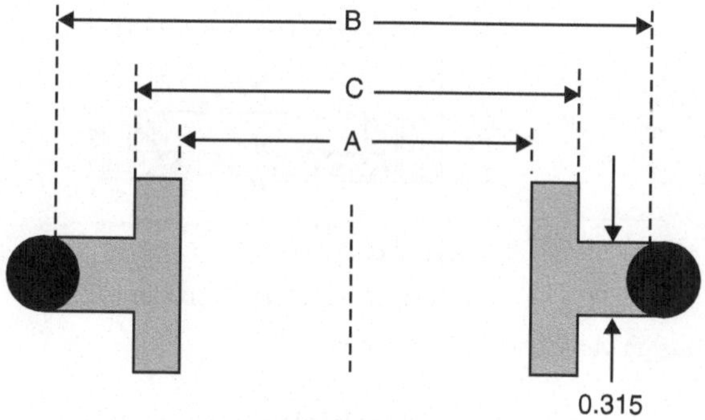

Centring Ring
All dimensions in inches

Table A.3: All dimensions in inches

NW	*A*	*B*	*C*
10-16	0.35	0.67	0.47
25	0.81	1.02	0.87
32-40	1.26	1.61	1.34

Figure A.7 and Table A.3: Centring ring for NW type flanges along with dimensions

Figure A.7 shows the centering ring with O-ring groove on which an O-ring is fixed. Table A.3 gives the dimensions for various sizes. The sizes of different O-rings are given in Table A.4. The clamps are in two halves and are such that they can be easily assembled on the flanges.

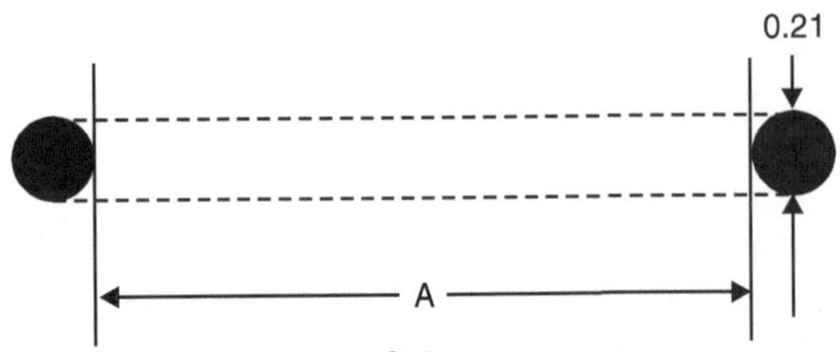

O-ring
All dimensions in inches

Table A.4:
All dimensions in inches

NW size	*A*
10	0.60
16	0.725
25	1.10
40	1.60
50	2.10

Figure A.8 and Table A.4: O-rings for NW type flanges along with dimensions

For the O-ring seal to work the sealing material must be deformable. It should have a low vapour pressure, a low out-gassing rate and a low permeability for diffusion of gas through it. Buna *N* O-rings are of low cost compatible with most chemicals found in vacuum applications. They can be used up to 80 C. Viton O-rings can stand upto 150 C. Silicone O-rings can be used for long periods upto 200 C. But silicone rubber hardens and sticks to the flange surfaces. These materials become brittle at low temperatures. If one wants a leak tight seal at low temperatures below 77 K, a good sealing material is Indium. This is a soft metal and is available in the form of wires. The wire is placed in the groove so that the two ends overlap. On bolting the flanges the metal flows easily and fills the groove.

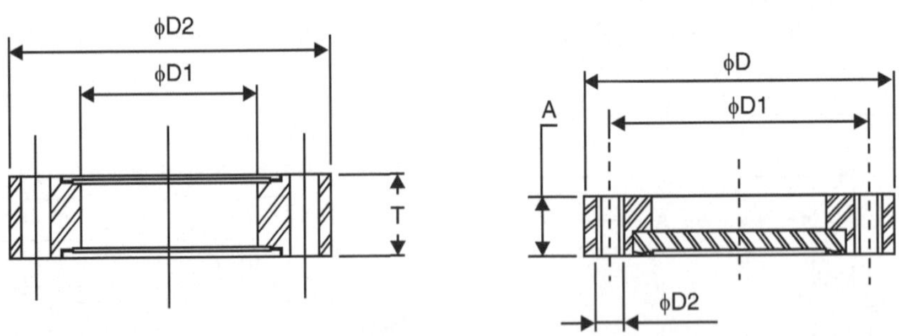

Bored double sided flanges

Blank flanges rotatable and tapped

Figure A.9: Design of typical Conflat flanges

The rubber O-ring material is not suitable for ultrahigh vacuum applications because of the relatively large out-gassing rate of these materials. In UHV it is customary to bake the vacuum system at a high temperature (200°C). These materials cannot withstand such baking. One then uses metal gaskets. The gasket will have to be soft so that it will flow on the application of pressure. The gasket material is usually oxygen-free high conductivity copper and the flanges are equipped with knife-edges between which the copper gasket is kept. For a given mechanical force on the flange, the pressure on the gasket material increases because of the small thickness of the knife edge and makes the metal to flow. Such a flange is called a Conflat flange. It is shown in Figure A.9.

8. VALVES, FEED-THROUGHS ETC.

A valve is used to isolate the vacuum chamber and preserve the low pressure in the chamber. In the high vacuum range one commonly uses diaphragm valves with the diaphragm made of special rubber material.

A feed-through is used to take out electrical connections from the vacuum chamber. In the high vacuum range a cost effective way of making these feed-throughs is by embedding metal rods of 1 to 2 mm diameter in a teflon disc for insulation. This teflon disc can be fitted to a metal flange with a suitable adhesive like Stycast.

For coupling a vacuum chamber and a tube one uses the following arrangement. The coupling has an internal O-ring seal which is compressed on the tube OD by finger tightening a knurled nut. This is shown in Figure A.10. When not in use, the coupling is blocked by solid plugs.

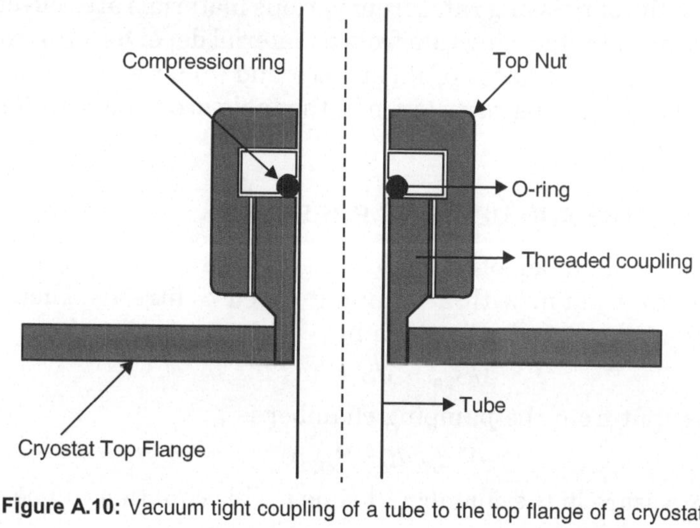

Figure A.10: Vacuum tight coupling of a tube to the top flange of a cryostat

9. OUTGASSING

In high and ultrahigh vacuum the liberation of gases from surfaces of materials plays an important role. Often the ultimate pressure achieved with a vacuum system is determined, not so much by fine leaks through the seals, but by the outgassing rates from the walls of the vessel.

Table A.5: Outgassing rates of some commonly used Materials

Material	*Condition*	*Outgassing rate in 10^{-8} std mbar 1/s/cm^2*		
		1 hr	*10 hrs*	*100 hrs*
Mild Steel	Shot-blasted		7.8	
Stainless Steel	degreased	170		
	polished	50	0.18	
	baked for 24 hrs at 200 C		0.02	
	baked for 12 hrs at 400 C		1.2×10^{-6}	
Aluminium	Anodized		13.0	
Aluminium	As received		0.3	
6061-T6	baked for 15 hrs at 300 C		0.05	
Copper	As received	330		
Brass Cast	Washed		4.0	
Mylar	Outgassed	26		
Neoprene	As received	20,000		
Teflon	As received	50	27	12
Kel-F	As received	4		
Polystyrene	As received	34		
Nylon	As received	1600		

In Table A.5 the outgassing rates from various materials are collected. It should be mentioned that the outgassing rate from a material depends on (*a*) the cleanliness of its surface, (*b*) the roughness of its surface and (*c*) on whether the material is baked or not. The outgassing rates given in the table are to be therefore considered as representative.

10. PUMPDOWN TIME AND ULTIMATE PRESSURE

Let us consider a chamber of volume V litres connected by a line of effective conductance C to a pump with a pumping speed S litres/s. Then the effective pumping speed at the chamber is given by

$$1/S_{eff} = 1/S + 1/C \tag{A.10.1}$$

The throughput from the pumping chamber is

$$-V\, dp/dt = S_{eff}\, p \tag{A.10.2}$$

Here p is the pressure in the chamber at time t. This equation can be solved to give

$$p(t) = p(0) \exp(-S_{eff}\, t/V) \tag{A.10.3}$$

In deriving the above equation we assumed that there is no outgassing from the walls of the chamber and the effective pumping speed is constant. Outgassing is unimportant in the pressure range from 1 bar to 100 millibar. Below this pressure outgassing gains in importance the lower the pressure. In the high vacuum range the throughput is mainly determined by outgassing from the walls of the chamber. Even in the pressure range 1 bar to 100 millibar one cannot use the above equation to determine the pump-down time since the effective pumping speed will not remain a constant. Experience has shown that the pumping time can be calculated from the equation

$$t = (KV/S_{eff}) \ln (p_i/p_f) \quad (A.10.4)$$

Here p_i is the initial pressure, p_f the final pressure and K is a numerical constant, which depends on the pressure range in which the equation is used.

Pressure range	**K**
1000 to 100 mbar	1.1
100 to 10 mbar	1.5
10 to 1 mbar	4.0

As an example let us work out the time taken to pump a volume *V* of 100 litres from 1 bar to 1 mbar using a pump of pumping speed 20 litres/s. Let the conductance of the tube be 10 litres/s. The effective pumping speed is 6.7 l/s. The time to pump from 1000 millibars to 100 millibars is t_1 = 38 s. The time to pump the chamber from 100 mbar to 10 mbar is t_2 = 52 s. The time to pump the chamber from 10 to 1 millibar is t_3 = 141 s. The total time required to pump the chamber from 1000 mbar to 1 millibar is 231 s or roughly 4 minutes.

On the other hand the calculation of the pump-down time below 1 millibar and the ultimate pressure that will be reached are determined largely by the outgassing rates from the walls of the chamber.

Let us consider another example in which the chamber is made of stainless steel with a surface area of 1000 cm^2. Let the chamber be pumped by a 100 l/s diffusion pump. Let the conductance of the line in this pressure range be 50 l/s. Then the effective pumping speed at the chamber is 33 l/s. From the Table A.5, the outgassing rate of degreased *SS* is 50×10^{-8} std. millibar l/s cm^2. So the total outgassing from the walls will be 5×10^{-4} std. millibar l/s. The rate at which pressure will decrease is given by

$$-V\, dp/dt + O = S_{eff}\, p \quad (A.10.5)$$

where *O* is the rate of outgassing from the walls of the chamber which is assumed to be independent of pressure. The ultimate equilibrium pressure that will be obtained in the system is

$$p_f = O/S_{eff} = 1.5 \times 10^{-5} \text{ mbar} \quad (A.10.6)$$

However long one waits the pressure will never go below this value. If we want to attain a lower pressure we have to reduce the out-gassing rate. This can be done by baking the stainless steel chamber to 200 C for 24 hours by winding a heating tape around it and running only the rotary pump. One must take care to see that the temperature at the locations where the elastomer seals are present does not go beyond the recommended value for the seal material. After baking, the outgassing rate reduces to 2×10^{-10} std. millibar l/s/cm^2 so that the total outgassing rate becomes $O = 2 \times 10^{-7}$ std millibar litre/sec. Now one may reach a pressure below 10^{-5} millibar. In fact the ultimate pressure that one can reach in the system will be 4×10^{-9} millibar. This shows the value of baking the chamber.

11. LEAK TESTING

A leak will reduce the ultimate pressure that can be reached in the vessel with a pump. If a valve is used to isolate the vacuum chamber from the pump after attaining an ultimate low pressure p_0, the pressure in the chamber will start increasing with time. This increase in pressure can be either due to a leak in the system or due to outgassing from the walls of the chamber. The leak occurs usually from atmosphere, the pressure of which is nearly 1 bar. The leak rate is proportional to the difference between atmospheric pressure and the pressure in the chamber. As long as the pressure in the chamber is below (say) 1 millibar, we may take the leak rate to be constant independent of the pressure in the chamber. On the other hand the outgassing rate from the walls of the chamber decreases rapidly as the pressure in the chamber increases.

Let us assume that the leak rate is Q std-cc/s. If the volume of the chamber is V in litres, then

$$1000\ V dp/dt = Q \tag{A.11.1}$$

or

$$(p(t) - p(0)) = (Q/1000\ V)\ t \text{ millibar} \tag{A.11.2}$$

The pressure rises linearly with temperature. One can estimate the leak rate Q by measuring the pressure rise as a function of time.

On the other hand if outgassing is responsible for the rise in pressure, the pressure p will saturate at a certain value (i.e.) the rate of rise in p will become less and less as the pressure increases. Usually outgassing will be appreciable only at pressures below about 10^{-5} millibar unless the walls of the chamber are very dirty.

If the leak rate is more than 10^{-4} std-cc/s we term it a gross leak. Otherwise it is called a fine leak. From the rise in pressure with time one may estimate if the leak is gross or fine. The estimated leak rate is the sum total of all leaks occurring at various places in the vacuum system.

A common procedure for testing gross leaks is to brush acetone at the suspected locations of the leak. Acetone has a high vapour pressure. If there is a gross leak the acetone vapour gets into the chamber and causes the pressure on the ionisation or thermocouple gauge to increase rapidly.

The best leak detection technique uses a mass spectrometer to detect helium leaking into the chamber. The detector has its own turbo-pump that evacuates the mass spectrometer to a high vacuum. If any helium enters the chamber it is sucked into the detector. The helium gas is ionised by an electron beam. The ionised helium then traverses through the mass spectrometer. The conditions in the mass spectrometer are set to detect only helium ions reaching the detector.

The nitrogen and oxygen atoms, which enter the leak detector and are ionised, are not detected. This gives a very low background signal in the detector, as the amount of helium commonly present in the atmosphere is very low. The mass spectrometer is connected to the vacuum chamber. If the chamber is small, the same turbo pump, which evacuates the mass spectrometer, can also be used to evacuate the chamber through a set of valves that are operated sequentially by microprocessor control in the leak detector.

The leak detector has in-built controls which will connect the vacuum chamber to the mass spectrometer, which is already evacuated to a low vacuum ($<10^{-8}$ mbar), only when the pressure in the chamber falls below a certain value (say 10^{-4} millibar). If the chamber is big, one can use an auxiliary pump of suitable speed to evacuate it. Helium gas from a cylinder with a low pressure regulator is then sprayed on the suspected location of the leak. The gas enters the mass spectrometer and its presence is signalled by a visual indication of the leak rate. If the leak rate exceeds a pre-set value, there is an audio signal to alert the investigator. Using an in-built calibrated leak (i.e. a radioactive alpha emitting material with a specified half life) one can periodically check the calibration on the visual indicator.

Commercial leak detectors, which operate with a single control switch, are available to detect leak rates as small as 10^{-10} std. cm^3/s. Such leak detectors are routinely used with systems where one would like to reach a pressure below 10^{-7} millibar.

In the case of pressure vessels leak detection can be done using the mass spectrometer leak detector in a sniffer mode. The pressure vessel is filled to a pressure of a few bars with air mixed with helium. A sniffer probe attached to the leak detector is used to suck the air-helium mixture leaking out of the chamber into the mass spectrometer leak detector.

Once the leak is detected, steps must be taken to close the leak. This may involve replacing a defective O-ring (or a copper gasket in the case of UHV systems). If there is a crack in the wall of the chamber or glass windows in the chamber, this will need to be corrected.

If the pressure in a chamber saturates some time after isolating it from the vacuum pump, outgassing is the possible cause. Outgassing can be reduced by baking the chamber to a temperature of about 200°C. The pump is kept on while the baking takes place. During baking, the pressure will first rise as the high temperature will increase the outgassing rate. But when all the gas has come out, the outgassing rate will go down and the pressure will start decreasing. One should wait till the pressure in the chamber becomes steady. The heater is then switched off and the chamber is allowed to cool to room temperature.

REFERENCES

1. Randall F. Barron, "Cryogenic Systems", 2nd edition, Oxford University Press, New York, 1985.
2. R.A. Haefer, "Kryo-Vakuumtechnik", Springer-verlag, Berlin, 1981.
3. A. Roth, "Vacuum Technology", North-Holland, Amsterdam, 1976.
4. M. Wutz, "Theorie and Praxis der Vacuumtechnik", Vierweg, Braunschweig, 1979.
5. Lewin G., "Fundamentals of Vacuum Science and Technology", McGraw Hill, New York (1965).
6. K. Diels, R. Jaeckel, "Laybold Vacuum-Taschenbuch", 2nd Edition, Springer verlag, Berlin (1962).
7. Varian Inc., Vacuum Technologies, Catalogue of vacuum components, website: http://www.varianinc.com/nav/products/vacuum/components/in.
8. John F. O'Hanlon, "User's guide to Vaccum Technology", John Wiley & Sons Inc., 1989.

Appendix A1 Design of Cryostats and Cryogenic Vessels

1. INTRODUCTION

A cryostat used for experiments is generally cylindrical in shape. In view of cryogenic temperatures, the cryostat will be double walled in construction, with the inner chamber subjected to the cryogenic temperatures. A cryogenic storage vessel is also similar in construction. In this section, we describe the design of a typical medium size storage container. Figure A1.1, shows the elements of the dewar vessel. The inner vessel is the product container, which stores the fluid. The inner vessel is surrounded by an outer vessel or vacuum jacket. The inter space is filled with suitable insulation and also evacuated to minimize the heat transfer from the outer to inner vessel. In small laboratory dewars, the insulation consists of the silvered walls and high vacuum. However, insulations such as powders, fibrous materials or multilayer insulations are used in larger vessels.

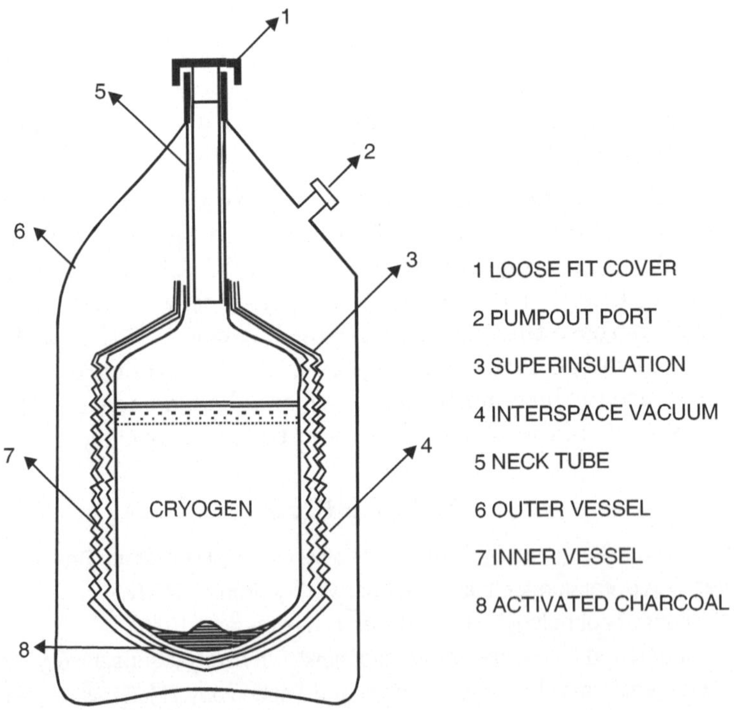

Figure A1.1: Schematic of a storage Dewar for Cryogenic Fluids

There is no need for fill and vent lines in smaller containers (the liquid is simply poured in or out through the open end of the container). However, fill and drain lines are necessary for larger containers. A vapour vent line must be provided to allow venting of vapour formed due to heat inleak. Liquid removal is accomplished by either pressurization of the inner vessel or by using a liquid pump. If pressurization is adopted, a vapour diffuser is incorporated in the vent line so that the gas is distributed in the ullage space (space above liquid). A suitable suspension system is required to support the product container within the outer vessel.

Cryogenic storage vessels may be constructed in a suitable shape such as spherical, cylindrical or conical etc. However, one of the most economic configurations is the cylindrical vessel with dished, elliptical or hemispherical heads, as end closures. Spherical shape is the most effective configuration as far as the heat inleak is concerned, but a cylindrical vessel with unit length to diameter ratio has only 21% greater surface area than the sphere of same volume, and so heat inleak is only marginally higher.

The design of cryogenic storage vessels are covered in standards such as the ASME Boiler and Pressure vessel code section VIII (1983) and the British Standards Institution Standard 1500 or 1515.

Inner Vessel Design

According to the ASME code, the minimum thickness of inner shell for a cylindrical vessel should be determined from,

$$t = \frac{p \times D}{2 \times S_a \times e_w - 1.2 \times p} = \frac{p \times D_o}{2 \times S_a \times e_w + 0.8 \times p} \tag{A1.1}$$

Here, p is the design pressure (absolute pressure for vacuum jacketed vessel), D is the inner diameter of shell, D_o is the outer diameter of shell, S_a is the allowable stress (approximately equal to one fourth of ultimate strength of material) and e_w is the weld efficiency.

Values of allowable stresses for cryogenic materials may be found in the literature given at the end of this chapter. The minimum thickness for spherical shells, hemispherical heads or torispherical heads is determined from,

$$t_h = \frac{p \times D \times K}{2 \times S_a \times e_w - 0.2 \times p} = \frac{p \times D_o \times K}{2 \times S_a \times e_w + 2 \times p \times (K - 0.1)} \tag{A1.2}$$

Here, D is the inner diameter for spherical vessel or hemispherical head, or the inside major diameter.

The value of constant K is given by,

$$K = 1/6 \times [2 + (D/D_1)^2] \tag{A1.3}$$

Where D_1 is the minor diameter of elliptical head.

Outer Vessel Design

The outer vessel is subjected to atmospheric pressure on one side and vacuum at the other. Due to this, the outer shell would not fail due to excessive stress, but due to elastic instability (collapsing or bucking). A thin cylindrical shell will collapse under an externally applied pressure at stresses much lower than the yield strength for the material. The collapsing pressure for a long cylinder is given by,

$$p_c = \frac{2 \times E \times (t/D_o)^3}{1-\gamma^3} \tag{A1.4}$$

Here, E is the Young's modulus of the shell material, t is the shell thickness, D_0 is the outer diameter and γ is the Poisson's ratio. Similarly, the collapsing pressure for a short cylinder is given as,

$$p_e = \frac{2.42 \times E \times (t/D_o)^{5/2}}{(1-\gamma^2)^{3/4} \times [L/D_o - 0.45 \times (t/D_o)^{1/2}]} \tag{A1.5}$$

L is the length of the cylinder. A cylinder is defined as long, when its length to diameter ratio meets the following condition,

$$L/D_o > 1.140 \times (1-\gamma^2)^{1/4} \times (D_o/t)^{1/2} \tag{A1.6}$$

Otherwise it is a short cylinder. The heads of the outer vessel must also withstand the collapsing load of the atmospheric pressure. The critical pressure for a hemispherical, elliptical or torispherical head is given by,

$$p_c = \frac{0.5 \times E \times (t_h/R_o)^2}{[3 \times (1-\gamma^2)]^{1/2}} \tag{A1.7}$$

Here, R_0 is the radius of the spherical head, or equivalent radius of elliptical head or crown radius of torispherical head.

The equivalent radius for elliptical heads is given by, $R_0 = K_t \times D$, where, D is the major diameter and K_t is a factor, which depends on the ratio of major to minor diameters of the ellipse. The collapsing or critical pressure as per ASME code is given by,

$$p_c = 4 \times p_a \tag{A1.8}$$

p_a is the allowable external pressure (equal to atmospheric pressure for the outer vessel of a storage vessel). More details with regard to the design of stiffening rings for inner and outer vessels and also the support structures are given in detail in Barron [1]. Enclosed at the end of this chapter is a simple program written in Fortran 90 for the calculation of wall thicknesses of cylindrical shells and dished ends based on the ASME pressure vessel code section VIII (unfired pressure vessels).

The following gives a FORTRAN Program for the determination of the wall thicknesses of the Cryogenic or Vacuum vessels based on the ASME Pressure Vessel Code section VIII (Unfired Pressure Vessels).

```
      ************************************************************************
---------------------------------------------------------------------------------

      Program CRYODESN

****************************************************************************
   c This program calculates the wall thickness of cylindrical
vessels
   c based on the bursting and collapsing pressures
   c inputs: material of vessel, outer diameter, height of vessel
   c outputs: 1. wall thickness based on collapsible pressure
   c         2. wall thickness based on bursting pressure
   c         3. volume of the vessel.
   c E is the Youngs modulus in units of Kg/mm2
     implicit double precision (a-h, o-z)
     character*15 Name, Name1
     common |Q| alpha, beta
     external f
     write(*,*)'WELCOME TO CRYOGENIC CONTAINER DESIGN'
     write(*,*)'The calculations are based on ASME code section
VIII'
     write(*,*)
  15 write(*,*)'Is the material of construction Al or SS ?'
     write(*,*)'Input "Al" for Aluminium'
     write(*,*)'Input "SS" for Stainless Steel'
     write(*,*)'Input "Cu" for Copper'
     read(*, 10) name
     if(name.eq.'Al')name1 = 'ALUMINIUM'
     if(name.eq.'SS')name1 = 'STAINLESS STEEL'
     if(name.eq.'Cu')name1 = 'COPPER'
  10  format (A10)
     if(Name.ne.'Al' .and. name.ne.'SS' .and. name.ne.'Cu')
     1goto 15
     if(Name.eq.'Al') E = 0.70e4
     if(Name.EQ.'SS') E = 2.1e4
     if(name.eq.'Cu') E = 1.17e4
     write(*,*)'input the OUTER DIAMETER in mm'
     read(*,*)Do
     write(*,*)'input the HEIGHT in mm'
     read(*,*)HT
     if(name.eq.'Al') gamma = 0.34
```

```
      if(name.eq.'SS') gamma = 0.26
      if(name.eq.'Cu') gamma = 0.33
c
c To find whether it is a LONG or SHORT cylinder
c
      write(*,*)'input the ROUGH WALL THICKNESS (mm) to be used'
      read(*,*)TTR
      write(*,*)
      LS = HT/DO
      RS = (1.537*(1.0-Gamma*Gamma)**0.25)/(TTR/DO)**0.5
      write(*,*)'LS,RS =',LS,RS
      write(*,*)
      if(LS.GT.RS) then
      write(*,*)'LS > RS ; So it is a LONG CYLINDER'
      else if(LS.LE.RS) then
      write(*,*)'LS < RS ; So it is a SHORT CYLINDER'
      write(*,*)
      endif
c
c calculation of thickness based on Collapsing pressure
c
      Write(*,*) 'WALL THICKNESS based on Collapsing pressure'
      write(*,*)
      write(*,*) 'Collapsing pressure assumed as 5 Kg/cm^2'
      write(*,*)
      pcoll = 5.0e-2
      if(LS.GT.RS) goto 100
c
c Calculation of coefficients
c
      Alpha = 0.45*pcoll*(1.0-Gamma*Gamma)**0.75/(2.42*E)
      Beta= pcoll*(1.0-Gamma*Gamma)**0.75*HT/(2.42*E*DO)
      A = 1.0
      B = 1.0e-5
      eta = 1.0e-20
      eps = 1.0e-20
      Call Root01(A, B, eps,eta, f, x, ifail)
      fx = f(x)
      thick = x*x*Do
      write(*,*) 'Calculated wall thickness = ', thick,' mm'
      di = do-2.0*thick
      goto 200
100   continue
c
```

```
c Calculations for a LONG CYLINDER
c
  tdo3 = pcoll*(1.0-gamma*gamma)/(2.0*E)
  thick = (tdo3)**(1.0/3.0)*do
  write(*,*)
  write(*,*) 'Calculated wall thickness =', thick,' mm'
  di = do-2.0*thick
200 continue
c
c Thickness of dished ends based on Collapsing Pressure
c
  write(*,*)
  write(*,*)'Thickness of dished ends based on Collapsing
Pressure'
  write(*,*)'Assuming HEMISPHERICAL SHAPE'
  ak = 0.5
  ro = ak*di
  pcoll = 5.0e-2
  thro2 = (pcoll*(1.0-gamma*gamma)**0.5)*3.0**0.5/(0.5*E)
  th = thro2**0.5*ro
  write(*,*)
  write(*,*) 'thickness of dished end =',th, ' mm'
  tc = th
c
c Calculations based on Bursting Pressure
c
  write(*,*)
  write(*,*) 'Wall thickness based on Bursting Pressure'
  write(*,*)
  write(*,*) 'Input INTERNAL PRESSURE for the vessel in psi'
  read(*,*)pint
  if(name.eq.'Al') sa=2350.0
  if(name.eq.'SS') sa=18750.0
  if(name.eq.'Cu') sa=6700.0
  ew = 0.7
  write(*,*)
  write(*,*) 'Weld Efficiency is assumed as 0.7'
  thik = pint*do/(2.0*(sa*ew+0.4*pint))
  write(*,*)
  write(*,*)'Calculated wall thickness =', thik, 'mm'
  write(*,*)
c
c Thickness of dished ends based on Bursting Pressure
c
```

```
      write(*,*) 'Thickness of dished ends based on Bursting pressure'
      write(*,*) 'Assuming HEMISPHERICAL SHAPE'
      ak = 0.5
      th = pint*di*ak/(2.0*sa*ew-0.2*pint)
      write(*,*)
      write(*,*) 'Thickness of dished ends =', th, 'mm'
      tb = th
c
c Volume Calculations
c
      if(thick.gt.thik)t = thick
      if(thick.le.thik)t = thik
      di = do-2.0*t
      vol = (22.0/7.0)*di*di/4.0*ht
      vol = vol/1.0e6
      write(*,*)
      write(*,*) 'VOLUME of the cylinder is =', vol, 'Litres'
      write(*,*)
      write(*,*)
      write(*,*)    'DESIGN DATA'
      write(*,*)
      write(*,*) 'Material of Construction = ', name1
      write(*,*)
      write(*,*) 'Outer Diameter of the vessel =', do, 'mm'
      write(*,*)
      write(*,*) 'Wall thickness of cylinder'
      write(*,*) 'based on Collapsing Pressure =', thick, 'mm'
      write(*,*)
      write(*,*) 'Wall thickness of Dished end'
      write(*,*) 'based on Collapsing Pressure =', tc, 'mm'
      write(*,*)
      write(*,*) 'Wall thickness of cylinder'
      write(*,*) 'based on Bursting Pressure  =', thik, 'mm'
      write(*,*)
      write(*,*) 'Wall thickness of Dished end'
      write(*,*) 'based on Bursting Pressure  =', tb, 'mm'
      write(*,*)
      write(*,*) 'Inner Diameter        =', di, 'mm'
      write(*,*)
      write(*,*) 'Height of the cylinder    =', ht, 'mm'
      write(*,*)
      write(*,*) 'Volume of the vessel      =', vol, 'Litres'
      end
c
```

```
c
c
      Function F(t)
      Implicit double precision(a-h,o-z)
      common/q/alpha,beta
      f = t**5+alpha*t-beta
      end
c
c
      Subroutine ROOT01(a,b,eps,eta,f,x,ifail)
      Implicit double precision(a-h,o-z)
      external f
      k = 0
      idec = 1
      x1 = a
      xp = a
      x = a
      f1 = f(a)
      fp = f1
      if(abs(fp)-eta)1,1,2
    2 x2 = b
      xn = b
      f2 = f(b)
      fn = f2
      x = b
      if(abs(fn)-eta)1,1,3
    3  if(fn*fp)5,4,4
    5  if(fp)6,6,7
    6 xp = b
      xn = a
      temp = fp
      fp = fn
      fn = temp
    7 if(f2-f1)8,9,8
    8 x = x1-f1*(x2-x1)/(f2-f1)
      if((x-x1)*(x2-x))9,9,12
    9 idec = -idec
      if(idec)10,10,11
   10  x = xn-fn*(xn-fp)/(fn-fp)
      goto 12
   11  x = (xn+xp)/2.
      deltx = x-x2
   12   if(abs(x-x2)-eps)1,1,14
```

```
14 x1 = x2
   x2 = x
   f1 = f2
   f2 = f(x)
   if(abs(f2)-eta)1,1,15
15 if(f2)16,16,17
16 xn = x
   fn = f2
   goto 7
17 xp = x
   fp = x2
   goto 7
 1 ifail = 0
   return
 4 ifail = 1
   return
   end
```

INDEX

M

N

O

P

Q

R

S